Annals of Mathematics Studies
Number 218

Comparison Principles for General Potential Theories and PDEs

Marco Cirant
F. Reese Harvey
H. Blaine Lawson
Kevin R. Payne

PRINCETON UNIVERSITY PRESS
PRINCETON AND OXFORD
2023

Requests for permission to reproduce material from this work should be sent to permissions@press.princeton.edu

Published by Princeton University Press
41 William Street, Princeton, New Jersey 08540
99 Banbury Road, Oxford OX2 6JX

press.princeton.edu

ISBN 9780691243610
ISBN (pbk.) 9780691243627
ISBN (e-book) 9780691243641

British Library Cataloging-in-Publication Data is available

Editorial: Diana Gillooly and Kiran Pandey
Production Editorial: Nathan Carr
Jacket/Cover Design: Heather Hansen
Production: Lauren Reese
Publicity: William Pagdatoon

This book has been composed in LaTeX

10   9   8   7   6   5   4   3   2   1

Contents

Appendix   171

Preface

In recent years there has evolved a symbiotic and productive relationship between fully nonlinear partial differential equations and generalized potential theories. For example, research on the complex Monge–Ampère equation has historically had an important connection to complex pluripotential theory. This connection now extends quite generally to all weakly elliptic nonlinear equations, and this book is dedicated to some important aspects of this story.

In particular, one main purpose here is to prove *comparison principles* for nonlinear potential theories in $\mathbb{R}^n$ in a very straightforward manner from *duality* and *monotonicity*. We will also show how to deduce comparison principles for nonlinear differential operators—a program seemingly different from the first. However, we will marry these two points of view, for a wide variety of equations, under something called the *correspondence principle*.

In potential theory, one is given a *subequation* (constraint set) $\mathcal{F}$ on the 2-jets of a function. The subequation $\mathcal{F}$ determines the potential theory via its associated $\mathcal{F}$-subharmonic (and $\mathcal{F}$-superharmonic) functions, which compete in the comparison principle. The boundary of $\mathcal{F}$ gives a differential equation $F = 0$ for a *compatible* operator F whose upper level set $\{F \geq 0\}$ is the constraint set $\mathcal{F}$. There are many differential operators, suitably organized around $\mathcal{F}$, which give the same equation. So potential theory gives a great strengthening and simplification to the operator theory. Conversely, the set of operators associated to $\mathcal{F}$ can have much to say about the potential theory. We will return to this interplay below.

An object of central interest here is that of monotonicity, which explains and unifies much of the theory. A *monotonicity cone* is a set of jets that map the constraint set into itself under addition. We will always assume that the maximal monotonicity cone for a potential theory has interior. This is automatic for gradient-free equations, where monotonicity is simply the standard degenerate ellipticity (positivity) and properness (negativity) assumptions.

We show that for each such potential theory $\mathcal{F}$, there is an associated *canonical operator* F, defined on the entire 2-jet space and having all the desired properties. Furthermore, comparison holds for this F on any domain $\Omega \subset\subset \mathbb{R}^n$ which admits a C^2-strictly $\mathcal{M}$-subharmonic function, where $\mathcal{M}$ is a monotonicity cone subequation for $\mathcal{F}$. For example, for the potential theory corresponding to convex functions, the canonical operator is the minimal eigenvalue of D^2u in the C^2-case. The utility of the general comparison theorem is greatly enhanced by a detailed study of monotonicity cone subequations $\mathcal{M}$ which admit the needed strict subharmonic.

On the operator side there is an important dichotomy into the *unconstrained case* and the *constrained case*. In the constrained case, the operator must be restricted to a proper subset of 2-jet space in order to have the minimal monotonicity necessary. Our approach performs such restrictions in a general way (avoiding an ad hoc treatment) which applies simultaneously to all operators F compatible with a given subequation $\mathcal{F}$. The unconstrained case is best illustrated by the *canonical operators*, whereas the constrained case is best illustrated by *Gårding–Dirichlet operators*.

We point out that many examples are given throughout this text from both pure and applied mathematics, and also from theoretical physics.

As stated above, this treatise also aims to advance the *interplay between potential theory and operator theory*, an investigation begun in a trio of papers [45–47] published in 2009, that has grown into a wide-ranging investigation with many interesting and important avenues still to pursue. There are many opportunities for cross-fertilization and synergy between the potential theory and the operator theory.

The motivating principles include the following:

First, the conditions imposed on a constraint set $\mathcal{F}$ correspond to and encode structural conditions and properties satisfied by the operator F. For example, the monotonicity property of *positivity* of the constraint set $\mathcal{F}$ corresponds to (*degenerate*) *ellipticity* of a compatible operator F. Also, in the pure second-order case, the condition $0 \notin \operatorname{Int} \mathcal{F}$ on the constraint case characterizes when the maximum principle is satisfied by every operator compatible with $\mathcal{F}$ (see [55]).

Second, the subequation $\mathcal{F}$ "frees" a given PDE from any particular form of the operator F. Many different operators correspond to the same constraint set $\mathcal{F}$. This *gives many operators with the same solution set*—a fact which gives much more freedom in the analysis. This is an important point in the work of Krylov [77] on the general notion of ellipticity. Moreover, $\mathcal{F}$ "liberates" the user from needing an operator F to apply nonlinear elliptic potential theory.

Third, "forgetting" about the operator leads to interesting questions that at first glance might not seem important for operator theory and provides a "machine" for formulating new conjectures and theorems. For instance, taking one's cue from known results in pluripotential theory or convex analysis, one is led to seek generalizations in other potential-theoretic situations as well. In this way, one can find welcome surprises in the operator theory.

Fourth, along with a rich abundance of geometrically motivated potential theories, there are many new PDEs to discover. For example, while every calibrated geometry has an underlying potential theory, known "natural" smooth operators are rare gems.

Fifth, in the rare cases where a polynomial, or a well-known smooth function F in the jet variables, is known for a fixed $\mathcal{F}$-potential theory, the operator F will have much to say about the potential theory. For example, with such an operator one can take derivatives of the equation F. This is often very productive.

Sixth, while we will be focused primarily on the comparison principle in this book, an important new feature of the potential-theoretic approach is that it correctly identifies the boundary geometry of domains for which existence for the natural Dirichlet problem for $\mathcal{F}$-harmonic functions can be proven by Perron's method. This is discussed in Section 3 of the introduction. When combined with the correspondence principle, this geometric analysis carries over to the Dirichlet problem for all operators F which are compatible with the constraint set $\mathcal{F}$ defining the potential theory. This is a major accomplishment of the theory.

A deeper appreciation of the interplay between potential theory and operator theory benefits from a comparison with conventional operator-theoretic methods.

There is an existing and powerful theory of *viscosity subsolutions/super-solutions* for fully nonlinear PDEs initiated by Crandall–Lions [31] with the important foundational contributions of Jensen, Ishii, and Evans. The potential theory approach borrows from, but can also clarify and give new perspectives on, viscosity theory.

The correspondence principle builds a bridge between the two approaches. In the unconstrained case, $\mathcal{F}$-subharmonics/superharmonics are viscosity sub-solutions/supersolutions of the equation $F = 0$ for every compatible operator F satisfying $\mathcal{F} = \{F \geq 0\}$. However, in the constrained case, duality clarifies the appropriate notion of viscosity supersolutions (their negatives are dual $\mathcal{F}$-subharmonics) and leads to a general notion of *admissibility constraints*. This implements in a systematic way what is often done in an ad hoc way in the viscosity literature.

Some advantages and achievements of the potential-theoretic approach are outlined in the motivating principles above. In addition (for operators with variable coefficients), by combining monotonicity and duality with *fiberegularity*, the potential-theoretic approach had led to proofs of the comparison principle for operators F that do not obey the standard structural conditions required by the conventional viscosity theory (see [23–25]). A simple example of this achievement is recalled in Example 1.28. Another advantage to the potential theory approach is illustrated by its success in treating operators on manifolds, as will be discussed briefly below.

Additional reflections concerning the interplay between potential theory and operator theory are given in the last section of the introductory chapter.

A few remarks concerning the term *fully nonlinear* for a partial differential equation are in order. By this we mean a general PDE of the form

$$F(x, u, Du, D^2u) = 0,$$

where the dependence on the second-order derivatives may be nonlinear, but not necessarily. Our motivation for this choice is threefold. First, most (but not all) of the examples that we consider in the book have nonlinearities in the second-order derivatives. Second, the linear theory and the quasilinear theory (where there are no nonlinearites in second-order derivatives) are already quite well developed, while the general fully nonlinear theory is much less understood. Important references for the fully nonlinear theory can be found in the monographs [14, 78] and [41, Chapter 17]. Third, in this general context, pointwise control obtained by maximum and comparison principles has proven to be a robust tool and is the main technique employed in the book. Hence the use of fully nonlinear PDEs throughout the text.

For simplicity of the exposition and in order to make the discussion more accessible to analysts, we will focus on the Euclidean setting of open subsets X of $\mathbb{R}^n$, although X could also be a Riemannian manifold as in [49, 50], or an almost complex manifold as in [54]. We point out that the notion of *local affine jet equivalence*, introduced in [49], has been shown to be a powerful tool for questions on manifolds, but those results are also quite useful in $\mathbb{R}^n$. For example, the set of equations equivalent to constant-coefficient ones is quite extensive. We refer the reader to the papers above.

There are many recent papers which have picked up the major theme of this book while working on manifolds, including works which examine the interplay between potential theory and geometry via interesting differential operators on manifolds. Some of these will be mentioned in the last section of the introductory chapter.

Finally, we focus on the constant-coefficient case because, in this situation, monotonicity and duality alone suffice to prove comparison for compatible operator–subequation pairs $(F, \mathcal{F})$. Much more can be said when dependence on spatial coordinates is added into the pair, but additional conditions must be imposed in order to prove the comparison principle. In particular, the notion of *tameness* plays an important role in inhomogeneous equations $F(J^2 u) = \psi(x)$, while the *fiberegularity* mentioned above plays an important role in more general equations $F(x, J^2 u) = 0$. This is discussed in Section 1.8.

The work of Blaine Lawson was partially supported by the Simons Foundation and the work of Marco Cirant and Kevin Payne was partially supported by the GNAMPA-INdAM (Gruppo Nazionale per l'Analisi Matematica, la Probabilità e le loro Applicazioni - Istituto Nazionale di Alta Matematica). The authors gratefully acknowledge this support. Marco Cirant, Reese Harvey, and Blaine Lawson would like to express their gratitude to Kevin Payne who initiated this project, and has worked tirelessly and with artful leadership to bring it to conclusion.

The authors, June 2022

Guide for the Reader

Chapter and section titles have been chosen with some care in the hopes that interested readers can easily navigate their way through the book.

The book has four parts with distinct objectives.

Part I: A Comprehensive Introduction (Chapter 1) which includes

- the key concepts and highlights organized by the main topics;
- an extensive collection of relevant examples;
- statements of the major results and applications;
- comparisons with the literature and reflections on potential theory and operator theory, including a brief historical account.

Part II: The Potential Theory Approach (Chapters 2–10) which includes a detailed analysis of

- subequation constraints, their subharmonics, and duality;
- monotonicity cone subequations including the *fundamental family* and classification of those cones which admit the needed C^2-strict subharmonic for comparison;
- the *monotonicity duality method* for proving the comparison principle;
- numerous illustrations including improvements, limitations, and special cases (reductions).

Part III: Marrying Potential Theory to Operator Theory via the Correspondence Principle (Chapter 11) with a detailed analysis of

- compatible operator–subequation pairs;
- a structure theory for subequations from monotonicity;
- construction of canonical and Gårding–Dirichlet operators, which illustrate unconstrained and constrained cases.

Part IV: Applications to PDEs (Chapter 12) with a detailed analysis of many classes of degenerate elliptic operators with sufficient monotonicity including

- gradient-free operators (where comparison always holds on arbitrary domains);
- operators with some strict monotonicity and with gradient dependence (where comparison always holds, perhaps with limitations on domain size);
- parabolic operators.

In addition there are four appendices. The first two supplement the main text and the last two collect some elementary (but important) facts about $\mathcal{F}$-subharmonic functions that have been included for completeness.

Part I

A Comprehensive Introduction

Chapter One

A Comprehensive Introduction

One of the main purposes of this work is to prove *comparison principles* with respect to a constant-coefficient nonlinear potential theory, in a very straightforward manner, from *duality* and *monotonicity*. We will also show how to deduce comparison principles for nonlinear differential operators, a program which seems somewhat different from the first. However, we will marry these two points of view, for a wide variety of equations, under something we call the *correspondence principle*. This turns out to be interesting for several reasons. In potential theory, one is given a constraint set $\mathcal{F}$ on the 2-jets of a function, and the boundary of $\mathcal{F}$ gives a differential equation. There are many differential operators, suitably organized around $\mathcal{F}$, which give the same equation. So potential theory gives a great strengthening and simplification to the operator theory. Conversely, the set of operators associated to $\mathcal{F}$ can have much to say about the potential theory.

These comments are exemplified by the following two basic cases. Consider the constraint set $\mathcal{P} = \{D^2 u \geq 0\}$. The potential theory associated to $\mathcal{P}$ is the full theory of convex functions. The equation given by $\partial \mathcal{P}$ is the homogeneous Monge–Ampère equation. However, it is also the equation $\lambda_1(D^2 u) = 0$, where λ_1 is the first ordered eigenvalue, and there are many more examples. For the second case, consider the constraint set $\mathcal{P}_{\mathbb{C}} = \{D^2_{\mathbb{C}} u \geq 0\}$, where $D^2_{\mathbb{C}}$ is the complex Hessian. Here, the associated theory is the wide field of pluripotential theory. The operators include $\det(D^2_{\mathbb{C}} u)$, $\lambda_1(D^2_{\mathbb{C}} u)$, and many others.

One of the main objectives of the treatise is to bring together these two points of view for a large and important class of equations, where many new results have been established.

A motivation for this study comes from the following consideration (other motivations will be given below). For the Dirichlet problem (DP) on a bounded domain Ω in Euclidean space, it is proved in [49] that *existence always holds* in the constant-coefficient case[1] (assuming that Ω has a smooth C^2-boundary satisfying the appropriate strict boundary convexity conditions). This leaves uniqueness, which the *comparison principle*, that for u a subsolution and v a supersolution,

$$u \leq v \text{ on } \partial\Omega \;\Rightarrow\; u \leq v \text{ on } \Omega,$$

---

[1]The conclusion "existence always holds" is precisely defined in Theorem A.2.

obviously implies. Interestingly, in our constant-coefficient case, using the fact that existence always holds (see Theorem A.5) one can show that uniqueness and comparison are actually equivalent.

An object of central interest here is that of *monotonicity*. It is monotonicity that explains and unifies much of the theory. For each constraint set $\mathcal{F}$ there exists a maximal set $\mathcal{M}$ with the monotonicity property $\mathcal{F} + \mathcal{M} = \mathcal{F}$. In many interesting cases, $\mathcal{M}$ is itself a constraint set. In simpler cases, such as pure second-order equations, or gradient-free equations, monotonicity comes down to the standard degenerate ellipticity and negativity assumptions. To explain this in more detail we need some notation. (The reader should note the many examples below, starting with Example 1.7, which may illuminate the following sections.)

1.1   THE POTENTIAL THEORY SETTING

Set $\mathcal{J}^2 := \mathbb{R} \times \mathbb{R}^n \times \mathcal{S}(n)$, the space of 2-jets with standard jet coordinates (r, p, A), where $\mathcal{S}(n)$ is the space of symmetric $(n \times n)$-matrices with real entries, and consider a set $\mathcal{F} \subset \mathcal{J}^2$. Then $\mathcal{F}$ is called a *constant-coefficient subequation constraint set* (or simply *subequation*, or *constraint set*) if $\mathcal{F}$ is not $\emptyset$ or $\mathcal{S}(n)$, and

$$\mathcal{F} + \mathcal{P}_0 \subset \mathcal{F}, \quad \mathcal{F} + \mathcal{N}_0 \subset \mathcal{F}, \quad \text{and} \quad \mathcal{F} = \overline{\operatorname{Int} \mathcal{F}}, \tag{1.1}$$

where $\mathcal{P}_0 := \{0\} \times \{0\} \times \mathcal{P}$ and $\mathcal{N}_0 := \mathcal{N} \times \{0\} \times \{0\}$ in $\mathcal{J}^2 = \mathbb{R} \times \mathbb{R}^n \times \mathcal{S}(n)$, with

$$\mathcal{P} := \{A \in \mathcal{S}(n) : A \geq 0\} \quad \text{and} \quad \mathcal{N} := \{r \in \mathbb{R} : r \leq 0\}. \tag{1.2}$$

Associated to a constraint set $\mathcal{F}$ is its *dual* constraint set[2]

$$\widetilde{\mathcal{F}} := \sim \{- \operatorname{Int} \mathcal{F}\} = -\{\sim \operatorname{Int} \mathcal{F}\}. \tag{1.3}$$

Now, each constraint set $\mathcal{F}$ determines a potential theory of $\mathcal{F}$-*subharmonic functions*. A C^2-function u on an open subset $X \subset \mathbb{R}^n$ is $\mathcal{F}$-*subharmonic* on X if

$$J^2_{x_0} u := (u(x_0), Du(x_0), D^2 u(x_0)) \in \mathcal{F} \quad \text{for all } x_0 \in X. \tag{1.4}$$

Using viscosity theory, this condition can be transferred pointwise from the 2-jet $J^2_{x_0} u$ to the set of *upper test jets* (see Definition 1.3) by requiring

$$J^2_{x_0} \varphi \in \mathcal{F} \quad \text{for all upper test functions } \varphi \text{ for } u \text{ at } x_0 \in X, \tag{1.5}$$

---

[2]Throughout this book, $\operatorname{Int} \mathcal{F}$ is the interior of $\mathcal{F}$ and $\sim \mathcal{F} = \mathcal{J}^2 \setminus \mathcal{F}$ the complement of $\mathcal{F}$ with respect to $\mathcal{J}^2$.

thereby extending the notion of $\mathcal{F}$-subharmonic from C^2-functions to the space $\mathrm{USC}(X)$ of all upper-semicontinuous, $[-\infty, \infty)$-valued functions on X.

In addition to the notion of duality (1.3), the other fundamental concept for this work is monotonicity.

Definition 1.1. A *monotonicity cone* for a subequation $\mathcal{F}$ is a cone $\mathcal{M} \subset \mathcal{J}^2$ (with vertex at the origin) such that

$$\mathcal{F} + \mathcal{M} \subset \mathcal{F}, \tag{1.6}$$

and in this case we say that $\mathcal{F}$ *is* $\mathcal{M}$-*monotone*.

Note that since $\mathcal{M}$ contains the origin, the inclusion (1.6) is an equality $\mathcal{F} + \mathcal{M} = \mathcal{F}$.

Since $\mathcal{F}$ is a subequation, one can always enlarge a monotonicity cone to one where

$$\mathcal{M} \supset \mathcal{N} \times \{0\} \times \mathcal{P} \quad \text{and} \quad \mathcal{M} \text{ is a closed convex cone.} \tag{1.7}$$

In fact, the closed convex cone hull of a monotonicity cone is also a monotonicity cone. For each $\mathcal{F}$ there is a *maximal* monotonicity cone. Moreover, in this work we are interested in subequations $\mathcal{F}$ which have monotonicity cones $\mathcal{M}$ which satisfy (1.7) and

$$\mathrm{Int}\,\mathcal{M} \neq \emptyset, \tag{1.8}$$

so that $\mathcal{M}$ is itself a subequation, called a *monotonicity cone subequation*. To see this, note that for a closed convex cone $\mathcal{M}$, we have

$$\mathrm{Int}\,\mathcal{M} \neq \emptyset \;\Leftrightarrow\; \mathcal{M} = \overline{\mathrm{Int}\,\mathcal{M}}.$$

From this assumption (1.8), which holds for many constraint sets (including all second-order, in fact, all gradient-free subequations), many important things follow:

- the correspondence principle;
- comparison theorems;
- the existence of canonical operators;
- the existence of unique solutions to the Dirichlet problem;
- and much more; see below.

1.2 THE DIFFERENTIAL OPERATOR SETTING

We now address the companion setting of differential operators. There are two cases: the unconstrained case and the constrained case.

Definition 1.2. A *compatible operator–subequation pair* $(F, \mathcal{F})$ consists of either

the unconstrained case: $\mathcal{F} = \mathcal{J}^2$ and $F \in C(\mathcal{F})$

or

the constrained case: $\mathcal{F} \subset \mathcal{J}^2$ is a subequation and $F \in C(\mathcal{F})$, with the properties

$$\inf_{\mathcal{F}} F > -\infty \quad \text{and} \quad \partial \mathcal{F} = \big\{ J \in \mathcal{F} : F(J) = c_0 \big\}, \tag{1.9}$$

where $c_0 := \inf_{\mathcal{F}} F \in \mathbb{R}$.

In both cases, $F(\mathcal{F})$ is called the set of *admissible levels* of the pair.

Classical examples in the constrained case are the real and complex Monge–Ampère operators, where one assumes $A \geq 0$ and (respectively) $A_{\mathbb{C}} \geq 0$; that is, one restricts to the subequations $\mathcal{P}$ and $\mathcal{P}_{\mathbb{C}}$. More generally, the constrained case is well illustrated by *Gårding–Dirichlet operators*, as discussed in Section 1.6. On the other hand, the unconstrained case is well illustrated by the so-called *canonical operators* as discussed in Section 1.4.

Let $\mathcal{M} \subset \mathcal{J}^2$ be a convex cone with vertex at the origin. We say that a compatible operator–subequation pair $(F, \mathcal{F})$ is $\mathcal{M}$-*monotone* if $\mathcal{F}$ is $\mathcal{M}$-monotone and

$$F(J + J') \geq F(J) \quad \forall J \in \mathcal{F}, \ \forall J' \in \mathcal{M}. \tag{1.10}$$

If $\mathcal{M} \supset \mathcal{N} \times \{0\} \times \mathcal{P}$, then $(F, \mathcal{F})$ is called *proper elliptic*.[3] These are the only operators we consider, because of our focus on comparison.

Next, these proper elliptic operators are divided into two classes: those which are *topologically pathological*, meaning that, as a function on the 2-jet space, the operator has a level set with interior, and those which are referred to as *topologically tame*. The topologically pathological case is discarded here because uniqueness of solutions (and hence comparison) is trivially impossible. Therefore,

all proper elliptic operators in this book will be assumed to be tame.

Various equivalent formulations of topological tameness appear in Theorem 11.10. For this topologically tame case, we will establish a rigorous *correspondence principle* between potential theory and PDEs.

---

[3]Often in the viscosity literature (such as [30]), one uses the simpler term *proper* for the $\mathcal{N} \times \{0\} \times \mathcal{P}$-monotonicity, but we prefer the phrase *proper elliptic* to recall both the $\mathcal{P}$-monotonicity (degenerate ellipticity or positivity) and $\mathcal{N}$-monotonicity (properness or negativity) for us.

1.3   THE CORRESPONDENCE PRINCIPLE

This result builds a bridge between nonlinear potential theory (subharmonics for a subequation and its dual) and nonlinear PDEs (admissible viscosity sub/supersolutions of PDEs); in particular, it represents the part of the theory using monotonicity and duality for which the two approaches are equivalent.

The main tool will come from *viscosity theory*, which was developed by Crandall, Ishii, Lions, Jensen, Evans, and others (see [30] and the references therein). For what we have to say in this work, one should note that it was Jensen [69] who made the initial breakthrough on comparison principles for viscosity solutions of fully nonlinear PDEs. Additional remarks on the development of viscosity solution techniques will be given in Section 1.9.

One of the fundamental definitions from viscosity theory is the following.

Definition 1.3. Let $x_0 \in X \subset \mathbb{R}^n$ (an open subset) and $u \in \mathrm{USC}(X)$. An *upper test function* for u at x_0 is a C^2-function φ defined near x_0 with

$$u(x) \le \varphi(x) \quad \text{and} \quad u(x_0) = \varphi(x_0).$$

A *lower test function* for u at x_0 is a C^2-function φ such that $-\varphi$ is an upper test function for $-u$ at x_0. We will denote by $J^{2,\pm}_{x_0}u \subset \mathcal{J}^2$ the spaces of (*upper/lower*) *test jets for u at x_0*, that is, the set of all $J = J^2_{x_0}\varphi$ where φ is a C^2 (upper/lower) test function for u at x_0.

For compatible pairs $(F, \mathcal{F})$ which admit a monotonicity cone subequation, there is a potential theory at each admissible level $c \in F(\mathcal{F})$.

Definition 1.4. Let $(F, \mathcal{F})$ be a compatible operator–subequation pair, which admits a monotonicity subequation $\mathcal{M}$, and let $c \in F(\mathcal{F})$ be an admissible level. Consider the subequation $\mathcal{F}_c \equiv \{J \in \mathcal{F} : F(J) \ge c\}$. Let $u \in \mathrm{USC}(X)$, where $X \subset \mathbb{R}^n$ is an open subset. Then u is said to be $\mathcal{F}_c$-*subharmonic* on X if

$$J^{2,+}_{x_0}u \subset \mathcal{F}_c \quad \text{for all } x_0 \in X. \tag{1.11}$$

Let $\mathrm{LSC}(X)$ denote the space of lower-semicontinuous $(-\infty, +\infty]$-valued functions on X. A function $v \in \mathrm{LSC}(X)$ is said to be $\mathcal{F}_c$-*superharmonic* on X if $-v$ is $\widetilde{\mathcal{F}}_c$-subharmonic on X. By duality, this is equivalent to asking that $J^{2,-}_{x_0}v \subset {\sim}\,\mathrm{Int}\,\mathcal{F}_c$ for each $x_0 \in X$.

We now present an essential notion for the constrained case, that is, of admissible viscosity subsolutions and supersolutions where the subequation $\mathcal{F}$ places a constraint on the upper/lower test jets that compete in the definition. In particular, part (b) in the definition below makes systematic what is often done in an ad hoc way in the literature.

Definition 1.5. Let $(F, \mathcal{F})$ be a compatible operator–subequation pair as above. Let Ω be a domain in $\mathbb{R}^n$ and let $c \in F(\mathcal{F})$ be an admissible level.

(a) A function $u \in \text{USC}(\Omega)$ is said to be an $\mathcal{F}$-*admissible viscosity subsolution* of $F(u, Du, D^2u) = c$ *in* Ω if for every $x_0 \in \Omega$ one has

$$J \in J_{x_0}^{2,+}u \;\Rightarrow\; J \in \mathcal{F} \text{ and } F(J) \geq c. \tag{1.12}$$

(b) A function $w \in \text{LSC}(\Omega)$ is said to be an $\mathcal{F}$-*admissible viscosity supersolution* of $F(u, Du, D^2u) = c$ *in* Ω if

$$J \in J_{x_0}^{2,-}w \;\Rightarrow\; \text{either } [J \in \mathcal{F} \text{ and } F(J) \leq c] \text{ or } J \notin \mathcal{F}. \tag{1.13}$$

The following theorem is a main result of this monograph.

Theorem 11.13 (Correspondence principle for compatible pairs). *Let $(F, \mathcal{F})$ be a compatible proper elliptic operator–subequation pair, which is $\mathcal{M}$-monotone for a convex cone subequation $\mathcal{M}$. Suppose also that F is topologically tame. Let $c \in F(\mathcal{F})$ be an admissible value, and set $\mathcal{F}_c \equiv \{J \in \mathcal{F} : F(J) \geq c\}$ as above. Fix a domain $\Omega \subset \mathbb{R}^n$. Then,*

(a) *$u \in \text{USC}(\Omega)$ is an $\mathcal{F}$-admissible viscosity subsolution of $F(u, Du, D^2u) = c$ in Ω if and only if u is $\mathcal{F}_c$-subharmonic on Ω;*
(b) *$u \in \text{LSC}(\Omega)$ is an $\mathcal{F}$-admissible viscosity supersolution of $F(u, Du, D^2u) = c$ in Ω if and only if u is $\mathcal{F}_c$-superharmonic on Ω;*
(c) *comparison for the subequation $\mathcal{F}_c$ on a domain Ω is valid if and only if comparison for the equation $F(u, Du, D^2u) = c$ on Ω is valid.*

The correspondence principle is a very general and powerful tool, which needs to be "unpacked" in order to fully appreciate it. First, there is an important *dichotomy* between the *unconstrained case* $(F, \mathcal{J}^2)$ in which the operator F is proper elliptic on all of $\mathcal{J}^2$ and the *constrained case* $(F, \mathcal{F})$ where F is proper elliptic only when restricted to some compatible subequation $\mathcal{F}$. Note that in the constrained case, the constraint set $\mathcal{F}$ on the domain of the operator F is used in Definition 1.5 of $\mathcal{F}$-*admissible sub/supersolutions*, while in the unconstrained case, sub/supersolutions are in the standard viscosity sense.

Second, using this principle, one can reduce PDE comparison to potential-theoretic comparison, in order to free the operator from its particular form, retaining only the need to analyze its maximal monotonicity cone $\mathcal{M}$. This is done in Chapter 12 for many classes of operators.

A final (important) remark concerning the correspondence principle is in order.

Remark 1.6 (Correspondence and the Dirichlet problem). One important (and immediate) consequence of the correspondence principle is that there are two equivalent formulations of the Dirichlet problem; namely, given a bounded domain Ω in $\mathbb{R}^n$ and given $\varphi \in C(\partial\Omega)$, for each admissible value $c \in F(\mathcal{F})$ find $h \in C(\overline{\Omega})$ such that

$$h \text{ is } \mathcal{F}_c\text{-harmonic on } \Omega \quad \text{and} \quad h = \varphi \text{ on } \partial\Omega \tag{DP}$$

or

$$h \text{ is an } \mathcal{F}\text{-admissible solution of } F(J^2 h) = c \text{ on } \Omega \quad \text{and} \quad h = \varphi \text{ on } \partial\Omega. \tag{DP$'$}$$

The *potential-theoretic formulation* (DP) is equivalent to the *operator-theoretic formulation* (DP$'$) since the notion of "solution" in both formulations is that the corresponding (and equivalent) conditions (a) and (b) in Theorem 11.13 hold. For a more extensive discussion on this equivalence in a more general setting, see [65, Section 1.3].

In both formulations, when the comparison principle holds, the existence of a (unique) solution on Ω for each fixed φ can be obtained by *Perron's method* (see Theorem A.2). In general, the domain will need to be *boundary pseudoconvex* in a suitable strict sense. A new feature of the potential-theoretic approach begun in [46] and extended in [49] is that the required *boundary geometry* is determined by the subequation $\mathcal{F}$ which defines the potential theory. Each subequation $\mathcal{F}$ determines an *asymptotic interior* $\overrightarrow{\mathcal{F}}$, which is a cone. A domain is said to be $\mathcal{F}$-*pseudoconvex* if $\partial\Omega$ admits a local defining function which is C^2-strictly $\overrightarrow{\mathcal{F}}$-subharmonic. This is a local question. This boundary geometry can be characterized in terms of the second fundamental form. These local defining functions serve as the needed barriers in Perron's method and existence holds provided that Ω is both $\mathcal{F}$-pseudoconvex and $\widetilde{\mathcal{F}}$-pseudoconvex. In many situations, there is no boundary geometry; that is, there are no geometric restrictions to impose on the boundary in order to have existence. For example, this happens in the *uniformly elliptic* case. On the other hand, when there is boundary geometry, the computations are interesting (see [50, Section 7]). Many subequations (and hence many operators) have the same boundary geometry.

Finally, when the correspondence principle holds, this boundary geometry analysis carries over to all operators F which are compatible with $\mathcal{F}$. This has been successfully exploited for variable coefficient proper elliptic operators in [23, 24].

1.4 CANONICAL OPERATORS

This collection of operators gives some of the best illustrations of the unconstrained case. The construction starts with a subequation $\mathcal{F}$ which admits a monotonicity subequation $\mathcal{M}$. One then chooses an element $J_0 \in \text{Int } \mathcal{M}$. Associated to this is a *canonical operator* $F \in C(\mathcal{J}^2)$, defined on all of $\mathcal{J}^2$, with very nice properties. It is canonically defined via the structure theorem

(Theorem 11.14), which says that for each $J \in \mathcal{J}^2$, the set

$$I_J := \{t \in \mathbb{R} : J + tJ_0 \in \mathcal{F}\} \tag{1.14}$$

is a closed interval of the form $[t_J, +\infty)$ with $t_J \in \mathbb{R}$ (finite). Moreover,

(a) $J + tJ_0 \notin \mathcal{F} \Leftrightarrow t < t_J$;
(b) $J + t_J J_0 \in \partial \mathcal{F}$;
(c) $J + tJ_0 \in \operatorname{Int} \mathcal{F} \Leftrightarrow t > t_J$.

The canonical operator $\mathcal{F} \colon \mathcal{J}^2 \to \mathbb{R}$ is then defined by

$$F(J) = -t_J \tag{1.15}$$

and it has the following properties: F decomposes $\mathcal{J}^2$ into three disjoint pieces

$$\partial \mathcal{F} = \{F(J) = 0\}, \quad \operatorname{Int} \mathcal{F} = \{F(J) > 0\}, \quad \text{and} \quad \mathcal{J}^2 \setminus \mathcal{F} = \{F(J) < 0\}, \tag{1.16}$$

and F is strictly increasing in the direction J_0. In fact,

$$F(J + tJ_0) = F(J) + t \quad \forall J \in \mathcal{J}^2, \ \forall t \in \mathbb{R}. \tag{1.17}$$

Furthermore, F is proper elliptic on $\mathcal{J}^2$ and, in fact, it is $\mathcal{M}$-monotone. It is also Lipschitz (see Propositions 11.17, 11.19, and 11.25).

Interestingly, there exist important cases where this construction is quite useful, that is, because there exist cases where there are probably no polynomial operators. Examples come from the geometrical potential theories for the special Lagrangian, G(2), and Spin(7). A different general construction of a natural operator to any subequation, namely the signed distance operator, is examined in Remark 1.10 below.

Concerning canonical operators, we have the following two results.

Theorem 11.20 (Canonical operators and compatible pairs). *Suppose that a subequation $\mathcal{F}$ admits a monotonicity cone subequation $\mathcal{M}$. Let $F \in C(\mathcal{J}^2)$ be the canonical operator for $\mathcal{F}$ determined by any fixed $J_0 \in \operatorname{Int} \mathcal{M}$. Then,*

(a) *$(F, \mathcal{J}^2)$ is an unconstrained proper elliptic operator–subequation pair;*
(b) *$F(\mathcal{J}^2) = \mathbb{R}$ and the operator F is topologically tame;*
(c) *for each $c \in \mathbb{R}$, the set $\mathcal{F}_c := \{J \in \mathcal{J}^2 : F(J) \geq c\}$ is a subequation constraint set with $\mathcal{F}_0 = \mathcal{F}$ and the pair $(F, \mathcal{F}_c)$ satisfies the compatibility conditions*

$$\inf_{\mathcal{F}_c} F = c \quad \text{and} \quad \partial \mathcal{F}_c = \{J \in \mathcal{F}_c : F(J) = c\}.$$

In addition, the canonical operator (determined by $J_0 \in \operatorname{Int} \mathcal{M}$) for the dual sub-equation $\widetilde{\mathcal{F}}$ is given by

$$\widetilde{F}(J) := -F(-J) \quad \text{for all } J \in \mathcal{J}^2.$$

Also note that

$$\widetilde{\mathcal{F}}_c = \mathcal{F}_{-c}.$$

Statements analogous to (a), (b), and (c) hold for $(\widetilde{F}, \mathcal{J}^2)$ and $(\widetilde{F}, \widetilde{\mathcal{F}}_c)$.

Theorem 11.21 (Comparison for canonical operators). *Let $\mathcal{F} \subset \mathcal{J}^2$ be a sub-equation constraint set which admits a monotonicity cone subequation $\mathcal{M}$. Further, suppose that $\mathcal{M}$ admits a strict approximator ψ on a bounded domain Ω, that is, $\psi \in C(\overline{\Omega}) \cap C^2(\Omega)$ such that $J_x^2 \psi \in \operatorname{Int} \mathcal{M}$ for each $x \in \Omega$. Then, for each $J_0 \in \operatorname{Int} \mathcal{M}$ fixed, the canonical operator F for $\mathcal{F}$ determined by J_0 satisfies the comparison principle at every level $c \in \mathbb{R}$, that is,*

$$u \leq w \text{ on } \partial\Omega \;\Rightarrow\; u \leq w \text{ on } \Omega$$

for $u \in \operatorname{USC}(\overline{\Omega})$ and $w \in \operatorname{LSC}(\overline{\Omega})$, which are respectively viscosity subsolutions and supersolutions to $F(u, Du, D^2u) = c$ on Ω.

This gives rise to many beautiful operator–subequation pairs, starting simply with just the subequation $\mathcal{F}$ itself.

Example 1.7 (Minimal eigenvalue operator). For the simplest example of a canonical operator, let $\mathcal{F} = \mathbb{R} \times \mathbb{R}^n \times \mathcal{P}$ (the real convexity subequation), and take $J_0 = (0, 0, \frac{1}{n}I)$. Then

$$F(J) = F(r, p, A) = \lambda_1(A) \quad \text{(the smallest eigenvalue of A).}$$

Of course, there are many other operators which are compatible with $\mathcal{F}$ and are zero on $\partial\mathcal{F}$, such as $\det(A)$ or $\det(A)^{\frac{1}{n}}$. However, for any such operator we know from Theorem 11.21 above that *comparison always holds*.

It is interesting to note that all linear operators are canonical (see Lemma 12.18). In addition, the concave operator F, which is the infimum over a suitably renormalized *pointed family* $\mathfrak{F} = \{F_\sigma\}_{\sigma \in \Sigma}$ of linear operators, is also the canonical operator for the convex cone subequation $\mathcal{F}$ which is the intersection of the associated half-space constraint sets $\{\mathcal{F}_\sigma\}_{\sigma \in \Sigma}$ (see Theorem 12.21). Similar considerations hold for the canonical supremum operator associated to the closure of the union of the $\mathcal{F}_\sigma$ (see Remark 12.31). The precise notion of being pointed is given in Definition 12.20 and is a geometrical hypothesis (see Remark 12.22) on the set of coefficient vectors

$$S = \left\{ J_\sigma = (a_\sigma, b_\sigma, E_\sigma) \right\}_{\sigma \in \Sigma} \subset \mathcal{J}^2$$

defining the operators in the family by

$$F_\sigma(J) = F_\sigma(r, p, A) := \operatorname{tr}(E_\sigma A) + \langle b_\sigma, p \rangle + a_\sigma r = \langle J_\sigma, J \rangle, \quad J \in \mathcal{J}^2. \qquad (1.18)$$

In the proper elliptic case, where each $(a_\sigma, E_\sigma) \in \mathcal{N} \times \mathcal{P}$, one also has the validity comparison principle (see Theorem 12.26) which depends on the interesting fact that a necessary and sufficient condition for the canonical operator for $\mathcal{F}$ to be $\mathcal{M}$-monotone is that S is contained in the convex polar cone $\mathcal{M}^\circ$ of $\mathcal{M}$. To facilitate the application of Theorem 12.26, the polars of many monotonicity cones are listed in Proposition 12.27. The following example application comes from optimal control and is discussed in Example 12.29.

Example 1.8 (Hamilton–Jacobi–Bellman operators with directed drift). In optimal control, one problem concerns an agent who seeks to minimize an infinite-horizon discounted cost functional by acting on its drift and volatility parameters. The relevant operator to consider is the infimum over a family of linear operators like (1.18), where we will specialize to

$$F_\sigma(J) = F_\sigma(r, p, A) = \operatorname{tr}(E_\sigma A) + \langle b_\sigma, p \rangle + cr = \langle J_\sigma, J \rangle, \quad \sigma \in \Sigma, \qquad (1.19)$$

where $\delta := -c > 0$ is the *discount factor*, b_σ is the *drift term*, and E_σ is the (*squared*) *volatility*. Under the assumptions that E_σ is allowed to vary in bounded sets and the set of drifts $S_d := \{b_\sigma\}_{\sigma \in \Sigma}$ share a "preferred" direction b_0 (the family is pointed with axis $b_0 \in \mathbb{R}^n \setminus \{0\}$), Theorem 12.26 shows that the comparison principle holds on arbitrary bounded domains for the equation $F(u, Du, D^2 u) = c$ for each $c \in \mathbb{R}$.

We now consider an important example of an unconstrained operator that is *not* a canonical operator. This particular equation has received much attention in recent years from quite varied points of view. There is some history in [62].

Example 1.9 (Special Lagrangian potential operator). This pure second-order operator was introduced along with special Lagrangian geometry in [44]. It takes the form

$$F(A) := \sum_{k=1}^{n} \arctan(\lambda_k(A))$$

and is $\mathcal{P}$-monotone on all of $\mathcal{J}^2$. Comparison on arbitrary bounded domains holds for the equation $F(D^2 u) = c$ at all admissible levels $c \in (-n\pi/2, n\pi/2)$, as shown in [46], which was a major motivation for that work. The constant c represents a (constant) *phase* (as described in [62]) and the problem is also interesting for nonconstant phases. The comparison principle in the nonconstant phase case has been completely settled in [13, 24].

We end this section with another class of operators naturally associated to any subequation.

Remark 1.10. An additional construction of a natural operator associated to any subequation constraint set $\mathcal{F} \subset \mathcal{J}^2$ is suggested by Krylov [77, Theorem 3.2] (in the pure second-order case). We like to call it the *(signed) distance operator*, which is defined by

$$\widehat{F}(J) := \begin{cases} \operatorname{dist}(J, \partial\mathcal{F}), & J \in \mathcal{F}, \\ -\operatorname{dist}(J, \partial\mathcal{F}), & J \in \mathcal{J}^2 \setminus \mathcal{F}. \end{cases} \tag{1.20}$$

The upper level sets of $\widehat{F}$ play a key role in the method for proving comparison in [49].

The signed distance operator $\widehat{F}$ is clearly Lipschitz and topologically tame (as is the canonical operator F) but is also well defined for any subequation $\mathcal{F}$ (while the canonical operator construction requires the $\mathcal{M}$-monotonicity of $\mathcal{F}$). On the other hand, when $\mathcal{F}$ is $\mathcal{M}$-monotone, the signed distance operator is *(linearly) tame* in the sense that it satisfies an inequality similar to (but weaker than) identity (1.17) satisfied by the canonical operator F. The tameness condition on an operator plays an important role in treating comparison for inhomogeneous equations, as will be discussed in Section 1.8. Given the importance of tameness, we record that fact in the following lemma. For clarity and simplicity, we restrict attention to a constant-coefficient pure second-order subequation $\mathcal{F} \subset \mathcal{S}(n)$.

Lemma 1.11 (Tameness of the signed distance operator). *Let $\mathcal{F} \subset \mathcal{S}(n)$ be a constant-coefficient pure second-order subequation. Then the signed distance operator*

$$\widehat{F}(A) := \begin{cases} \operatorname{dist}(A, \partial\mathcal{F}), & A \in \mathcal{F}, \\ -\operatorname{dist}(A, \partial\mathcal{F}), & A \in \mathcal{S}(n) \setminus \mathcal{F}. \end{cases} \tag{1.21}$$

satisfies

$$\widehat{F}(A + tI) - \widehat{F}(A) \geq t \quad \forall A \in \mathcal{S}(n), \ \forall t \in \mathbb{R}, \tag{1.22}$$

or equivalently,

$$\widehat{F}(A + P) - \widehat{F}(A) \geq t \quad \forall A \in \mathcal{S}(n), \ \forall t \in \mathbb{R}, \ \forall P \geq tI. \tag{1.23}$$

Proof. Clearly, (1.23) implies (1.22) by using $P = tI \geq tI$. The reverse implication follows by writing $P = tI + P'$ with $P' \geq 0$ and using the degenerate ellipticity of $\widehat{F}$.

Notice that (1.22) is trivial for $t = 0$ and it suffices to prove (1.22) for $A \in \mathcal{F}$ and $t > 0$ by exploiting duality where the signed distance operator $\widetilde{\widehat{F}}$ of the dual subequation $\widetilde{\mathcal{F}}$ satisfies

$$\widetilde{\widehat{F}}(A) = -\widehat{F}(-A) \quad \forall A \in \mathcal{S}(n), \tag{1.24}$$

and both vanish on their common boundary.

Condition (1.22) for $A \in \mathcal{F}$ and $t > 0$ can be restated in terms of metric balls in $\mathcal{S}(n)$ as

$$\text{for all } r > 0, \quad \text{if } B_r(A) \subset \mathcal{F} \text{ then } B_{r+t}(A + tI) \subset \mathcal{F}. \tag{1.25}$$

Now $B_{r+t}(A + tI) = B_{r+t}(tI) + A = B_r(tI) + A + B_t(tI) = B_r(A + tI) + B_t(tI)$. Hence, if $B_r(A) \subset \mathcal{F}$ then

$$B_{r+t}(A + tI) = (B_r(A) + tI) + B_t(tI) \in \mathcal{F} + \mathcal{P} \subset \mathcal{F},$$

which gives (1.25) by the positivity of $\mathcal{F}$. $\square$

Notice that inequality (1.22) for the signed distance operator $\widehat{F}$ is weaker than identity (1.17), $F(A + tI) = F(A) + t$, satisfied by the canonical operator.

1.5 GRADIENT-FREE OPERATORS

Given a subequation $\mathcal{F}$, our results apply if the maximal monotonicity cone of $\mathcal{F}$ has interior. However, notice that *this is true for every pure second-order subequation* $\mathcal{F} = \mathbb{R} \times \mathbb{R}^n \times \mathcal{F}_0$ since $\mathbb{R} \times \mathbb{R}^n \times \mathcal{P}$ is always a monotonicity subequation for $\mathcal{F}$. In fact, *this is true for every pure gradient-free subequation* $\mathcal{F}$, since $\mathcal{N} \times \mathbb{R}^n \times \mathcal{P}$ is always a monotonicity subequation for $\mathcal{F}$ by Definition 10.2 of *gradient-free*. We have the following result.

Theorem 12.2 (Comparison in the gradient-free case). *Suppose that $(F, \mathcal{F})$ is a compatible, gradient-free pair. Then for every bounded domain Ω and every $c \in F(\mathcal{F})$, one has the comparison principle*

$$u \leq w \text{ on } \partial\Omega \; \Rightarrow \; u \leq w \text{ on } \Omega$$

for $u \in \mathrm{USC}(\overline{\Omega})$ and $w \in \mathrm{LSC}(\overline{\Omega})$, where u is $\mathcal{F}_c$-subharmonic and w is $\mathcal{F}_c$-superharmonic (that is, $-w$ is $\widetilde{\mathcal{F}}_c$-subharmonic).

We have seen that the unconstrained case is best illustrated by canonical operators. The constrained case is best illustrated by operators involving Gårding hyperbolic polynomials, which we examine in the next section.

1.6   OPERATORS INVOLVING GÅRDING–DIRICHLET POLYNOMIALS

Gårding's theory [40] provides a unified approach to studying many of the most important subequations. The reader should look at Section 11.6 for more details and at [48, 51] for a modern self-contained treatment. An important study of fully nonlinear PDEs exploiting Gårding's theory is contained in the paper of Caffarelli–Nirenberg–Spruck [15]. A *Gårding-Dirichlet polynomial* is a homogeneous polynomial $\mathfrak{g}$ of degree m on $\mathcal{S}(n)$ with the following properties:

(1) (*I-hyperbolicity*). For each $A \in \mathcal{S}(n)$, the polynomial $p_A(t) \equiv \mathfrak{g}(tI + A)$ has all real roots. The negatives of these real roots are called the *Gårding I-eigenvalues* of A and up to permutation can be written in increasing order as $\lambda_1^{\mathfrak{g}}(A) \leq \lambda_2^{\mathfrak{g}}(A) \leq \cdots \leq \lambda_m^{\mathfrak{g}}(A)$.

(2) (*Positivity*). We assume $\mathfrak{g}(I) > 0$ and define the *Gårding cone* Γ to be the connected component of $\mathcal{S}(n) \setminus \{\mathfrak{g} = 0\}$ which contains the identity I. This is a convex cone (see Theorem 11.30), given by those A with $\lambda_1^{\mathfrak{g}}(A) > 0$. We assume the positivity property

$$\overline{\Gamma} + \mathcal{P} \subset \overline{\Gamma}, \tag{1.26}$$

which is equivalent to either $\overline{\Gamma} + \mathcal{P} = \overline{\Gamma}$ or $\mathcal{P} \subset \overline{\Gamma}$, since $\mathcal{P}$ contains the origin and $\overline{\Gamma}$ is a convex cone.

We normalize so that $\mathfrak{g}(I) = 1$. Then we have

$$\mathfrak{g}(tI + A) = \prod_{k=1}^{m} (t + \lambda_k^{\mathfrak{g}}(A)), \tag{1.27}$$

which when evaluated at $t = 0$ shows that each Gårding–Dirichlet operator is a *generalized Monge–Ampère operator*, where the Gårding I-eigenvalues of A take the place of the standard eigenvalues of A in the special case $\mathfrak{g} = \det$.

   If $\mathfrak{g}(A)$ is a Gårding–Dirichlet polynomial on $\mathcal{S}(n)$ with closed Gårding cone $\overline{\Gamma}$, then this gives rise to a pure second-order polynomial operator $\mathfrak{g}(D^2 u)$ constrained to the pure second-order subequation $\mathbb{R} \times \mathbb{R}^n \times \overline{\Gamma}$. Such operators are the subject of Section 11.6.

   Simple examples of Gårding–Dirichlet polynomials $\mathfrak{g}(A)$ are given by the kth elementary symmetric function (with $k = 1, \ldots, n$) of the I-eigenvalues of A (the so-called *Hessian equations*) as discussed in Examples 11.33(2). The case $k = n$ corresponds to the Monge–Ampère operator and the case $k = 1$ corresponds to the Laplacian. There are many more interesting examples, including the Lagrangian Monge–Ampère operator (see Examples 11.33(4)) and the *geometric k-convexity operator* (see Example 1.13 below). Moreover, each of these *universal examples* (defined in terms of the standard eigenvalues $\lambda_k(A)$) gives rise to a huge family of examples by simply replacing the standard eigenvalues $\lambda_k(A)$ by the Gårding I-eigenvalues $\lambda_k^{\mathfrak{g}}(A)$ of A for any Gårding I-hyperbolic polynomial $\mathfrak{g}$ on $\mathcal{S}(n)$. More precisely, we recall the following result. For the proof, see [51, Proposition 7.7].

Proposition 1.12 (Universally defined Gårding–Dirichlet operators). *Suppose that* $\mathfrak{p} = \mathfrak{p}(\lambda_1, \ldots, \lambda_m)$ *is a symmetric homogeneous polynomial of degree N in m variables with nonnegative coefficients, which is e-hyperbolic for* $e := (1, \ldots, 1) \in \mathbb{R}^m$ *and satisfies* $\mathfrak{p}(e) > 0$. *Denote its Gårding cone by* $E \subset \mathbb{R}^m$ *and its Gårding e-eigenvalues by* $\Lambda_1^{\mathfrak{p}}(\lambda), \ldots, \Lambda_N^{\mathfrak{p}}(\lambda)$. *Then* $\mathfrak{p}$ *induces a Gårding–Dirichlet operator* F *on* $\mathcal{S}(n)$ *of degree N for each Gårding–Dirichlet operator* $\mathfrak{g}$ *on* $\mathcal{S}(n)$ *of degree m defined by*

$$F(A) := \mathfrak{p}(\lambda_1^{\mathfrak{g}}(A), \ldots, \lambda_m^{\mathfrak{g}}(A)), \tag{1.28}$$

with Gårding cone $\Gamma_F = (\lambda^{\mathfrak{g}})^{-1}(E)$. *The eigenvalues of F are*

$$\Lambda_k^F(A) = \Lambda_k^{\mathfrak{p}}(\lambda_1^{\mathfrak{g}}(A), \ldots, \lambda_m^{\mathfrak{g}}(A)) = \Lambda_k^{\mathfrak{p}}(\lambda^{\mathfrak{g}}(A)), \quad k = 1, \ldots, N.$$

The reader is encouraged to consult [51] for more details.

Such polynomials $\mathfrak{p} = \mathfrak{p}(\lambda)$ are referred to as the "universal eigenvalue operators," and the cone $\overline{E}$ is called a "universal eigenvalue subequation." One can show that the coefficients of $\mathfrak{p}$ being nonnegative is equivalent to E being $\mathbb{R}_{>0}^m$-monotone.

Iterating the construction in Proposition 1.12, together with taking products (since products of Gårding polynomials are again Gårding), allows one to show that the family of Gårding–Dirichlet operators is huge.

There are many interesting equations which involve Gårding–Dirichlet polynomials $\mathfrak{g}(A)$. We now look at some of the examples.

Example 1.13 (*k*-plurisubharmonicity, the truncated Laplacian, and the geometric *k*-convexity operator). These examples were introduced in [46, subsection of Section 10] as part of the *geometric p-plurisubharmonic Dirichlet problem*. Here they illustrate the general fact that, given a Gårding polynomial, there are two natural operators: the Gårding operator defined directly by $\mathfrak{g}$ and the canonical operator for the Gårding cone $\overline{\Gamma}$ determined by $\mathfrak{g}$. We discuss an interpolation of operators between them. First, we define the potential theory, which is quite interesting. Fix an integer k, $1 \leq k \leq n$. A *k-plurisubharmonic function* is defined by requiring that its restriction to every affine k-plane is classically Laplacian subharmonic (or $\equiv -\infty$). The subequation $\mathcal{P}(k)$ is defined by requiring that $A \in \mathcal{S}(n)$ restricts, as a quadratic form, to have a positive trace on all affine k-planes. The k-plurisubharmonic functions are exactly the $\mathcal{P}(k)$-subharmonics.

The canonical operator is the *truncated Laplacian*

$$\Delta_{\mathcal{P}(k)}(A) \equiv \lambda_1(A) + \cdots + \lambda_k(A), \tag{1.29}$$

where $\lambda_1(A) \leq \cdots \leq \lambda_n(A)$ are the ordered eigenvalues of A. There is also a polynomial Gårding–Dirichlet operator

$$T_k(A) = \prod_{i_1 < \cdots < i_k} (\lambda_{i_1}(A) + \cdots + \lambda_{i_k}(A)), \qquad (1.30)$$

which we call the *geometric k-convexity operator*. This yields two compatible operator–subequation pairs using the canonical operator and a Gårding–Dirichlet operator, namely

$$(\Delta_{\mathcal{P}(k)}, \mathcal{P}(k)) \quad \text{and} \quad (T_k, \mathcal{P}(k)),$$

and yields a new interpolated sequence between the pairs $(\lambda_1, \mathcal{P})$ and $(\det, \mathcal{P})$ at the $k = 1$ end, and the identical pairs $(\Delta, \{\mathrm{tr} \geq 0\})$ and $(\Delta, \{\mathrm{tr} \geq 0\})$ at the $k = n$ end. The canonical operator has been studied in [7] where the terminology *truncated Laplacian* was introduced.

We point out that $\mathcal{P}(k)$-subharmonic functions restrict to be subharmonic on all k-dimensional minimal submanifolds [53].

Example 1.14. The reader might enjoy the article [58] where one has a full-blown Lagrangian plurisubharmonic potential theory, complete with an operator of "Monge–Ampère type" in Lagrangian geometry.

Example 1.15 (Branches of a Gårding–Dirichlet operator). In Section 11.7 we discuss the general notion of branches. A *branch* is a closed subset of $\mathcal{J}^2$ which is the boundary of a subequation. Given a Gårding–Dirichlet polynomial $\mathfrak{g}$ of degree m, there are m distinct branches

$$\Lambda_1^{\mathfrak{g}} \subset \Lambda_2^{\mathfrak{g}} \subset \cdots \subset \Lambda_m^{\mathfrak{g}}, \quad \text{where } \Lambda_k^{\mathfrak{g}} = \{\lambda_k^{\mathfrak{g}} \geq 0\}.$$

Our theory applies to all of these branches, because they are pure second order. (The only natural operator for these branches is the canonical operator $\lambda_k^{\mathfrak{g}}$ unless $k = 1$.)

Example 1.16 (Gradient-free operators with a Gårding–Dirichlet factor). Let $\mathfrak{g}(A)$ be a Gårding–Dirichlet polynomial as above, and let $h \in C((-\infty, 0])$ be nonnegative, nonincreasing, and with $h(r) = 0 \Leftrightarrow r = 0$. Consider the operator

$$F(r, p, A) = h(r)\mathfrak{g}(A). \qquad (1.31)$$

Restricting F to the subequation $\mathcal{F} = \mathcal{N} \times \mathbb{R}^n \times \overline{\Gamma}$ gives a compatible gradient-free pair, and hence comparison holds at every admissible level on every bounded domain.

An interesting special case comes from affine hyperbolic geometry, as presented in Example 12.6.

Example 1.17 (The hyperbolic affine sphere equation). The partial differential equation

$$\det(D^2 u) = \left(\frac{L}{u}\right)^{n+2}, \quad L \leq 0, \quad \text{that is, } (-r)^{n+2} \det(A) = (-L)^{n+2}, \quad (1.32)$$

arises in the study of *hyperbolic affine spheres* with mean curvature L, where $u < 0$ is convex and vanishes on the boundary of $\Omega \subset \mathbb{R}^n$ convex (see Cheng–Yau [22]). This equation is covered by the example above if one takes $\mathfrak{g}(A) = \det(A)$, and $h(r) = (-r)^{n+2}$ and $c = (-L)^{n+2} \geq 0$ are the admissible levels.

The next example illustrates a new construction in Section 11.6 (see Lemma 11.35) which produces a gradient-free Gårding–Dirichlet operator from a pure second-order Gårding–Dirichlet operator.

Example 1.18. For each Gårding–Dirichlet polynomial $\mathfrak{g}$ of degree m on $\mathcal{S}(n)$ with Gårding I-eigenvalues of A given by $\lambda_k^{\mathfrak{g}}(A)$, $k = 1, \ldots, m$, the operator

$$\mathfrak{h}(r, A) = \prod_{k=1}^{m} (\lambda_k^{\mathfrak{g}}(A) - r) = \mathfrak{g}(A - rI) \quad (1.33)$$

is a $(-\frac{1}{2}, \frac{1}{2}I)$-hyperbolic Gårding–Dirichlet polynomial of degree m on $\mathbb{R} \times \mathcal{S}(n)$ (normalized to have $\mathfrak{h}(-\frac{1}{2}, \frac{1}{2}I) = 1$) with Gårding eigenvalues $\lambda_k^{\mathfrak{h}}(A) = \lambda_k^{\mathfrak{g}}(A) - r$.

Now we consider an example with gradient dependence which requires an additional *directionality property* (D) with respect to a *directional cone* $\mathcal{D}$ (see Definition 2.2).

Example 1.19 (Example 1.16 with a directional cone). Let $\mathfrak{g}$ and h be as in Example 1.16 above, and consider a continuous $d \colon \mathcal{D} \to \mathbb{R}$, where $\mathcal{D} \subsetneq \mathbb{R}^n$ is a directional cone, with

$$d \geq 0 \text{ and } d(p) = 0 \iff p \in \partial\mathcal{D}, \quad (1.34)$$

$$d(p + q) \geq d(p) \quad \text{for each } p, q \in \mathcal{D}. \quad (1.35)$$

Then the operator

$$F(r, p, A) = h(r)d(p)\mathfrak{g}(A) \quad (1.36)$$

with restricted domain

$$\mathcal{F} = \mathcal{N} \times \mathcal{D} \times \overline{\Gamma} \quad (1.37)$$

defines a compatible $\mathcal{N} \times \mathcal{D} \times \mathcal{P}$-monotone operator–subequation pair $(F, \mathcal{F})$. Hence, comparison holds on arbitrary bounded domains at every admissible level of F.

Note: Examples of such pairs $(d, \mathcal{D})$ include

$$d(p) = p_n \text{ and } \mathcal{D} = \{(p', p_n) \in \mathbb{R}^n : p_n \geq 0\} \quad \text{(a half-space)}, \quad (1.38)$$

and, for $k \in \{1, \ldots, n\}$,

$$d(p) = \prod_{j=1}^{k} p_j \text{ and } \mathcal{D} = \{(p_1, \ldots, p_n) \in \mathbb{R}^n : p_j \geq 0 \text{ for each } j = 1, \ldots, k\}. \quad (1.39)$$

An interesting special case of Example 1.19 concerns parabolic operators, which are discussed in Section 12.6 in both the constrained (Theorem 12.38) and the unconstrained cases (Theorem 12.37).

Example 1.20 (Parabolic operators). In the case where the gradient pair $(d, \mathcal{D})$ is defined by (1.38), $h \equiv 1$, and $\mathfrak{g}(A)$ is replaced by $G(A')$ which depends only on $A' \in \mathcal{S}(n-1)$ (second-order derivatives only in the *spatial variables* $x' \in \mathbb{R}^{n-1}$), one has a fully nonlinear *parabolic* operator

$$F(r, p, A) := p_n G(r, A') \quad (1.40)$$

of the kind considered by Krylov in his extension of Alexandroff's methods to parabolic equations in [75]. The compatible subequation is described in formula (12.144) of this book.

Another interesting special case of Example 1.19 comes from a very particular form of optimal transport with quadratic cost, as presented in Example 12.34.

Example 1.21 (Potential equation for optimal transport with uniform source density and directed target density). Equations of the form

$$d(Du) \det(D^2 u) = c, \quad c \geq 0 \quad (1.41)$$

arise in the theory of optimal transport, under some restrictive assumptions. In general, there would be a function $f = f(x)$ in place of the constant c, where f represents the mass density in the source configuration and d represents the mass density of the target configuration (with the mass balance $\|f\|_{L^1} = \|d\|_{L^1}$). One seeks to transport the mass with density f onto the mass with density d at minimal transportation cost (which is quadratic with respect to transport distance). The solution of this minimization problem is given by the gradient of a convex function u, which turns out to be a generalized solution of equation (1.41). In the special case of uniform source density $f \equiv c$ and with target density d having some directionality, comparison principles can be obtained as a special case of Example 1.19 with $h(r) := 1$ and $\mathfrak{g}(A) := \det A$.

Thus we see that seemingly diverse equations can be established from a surprisingly *unified point of view*. It frees the theory from any particular form of the operator. Given a potential theory, that is, given a subequation constraint set $\mathcal{F}$, there are many natural choices for an associated operator. If $\mathcal{F}$ has *sufficient monotonicity*, that is, if $\mathcal{F}$ admits a monotonicity cone subequation $\mathcal{M}$ (that is, the maximal monotonicity cone has interior), there is always one choice that is "canonical," but for proving useful estimates, other choices may be better. For instance, a polynomial operator, if there is one, may be preferable. Restricting attention, as we do here, to the continuous version of the Dirichlet problem (DP), the correspondence principle enables a single potential theory/subequation result to be applied to all of the many compatible operators F associated to the subequation $\mathcal{F}$.

1.7 GENERAL POTENTIAL-THEORETIC COMPARISON THEOREMS

One of the important parts of this monograph is understanding convex cone subequations $\mathcal{M} \subset \mathcal{J}^2$ and the comparison results for subequations $\mathcal{F}$ which are $\mathcal{M}$-monotone.

By comparison results, we mean the validity of the comparison principle on bounded domains $\Omega \subset \mathbb{R}^n$, that is,

$$u \leq w \text{ on } \partial\Omega \implies u \leq w \text{ on } \Omega \tag{1.42}$$

for all $u \in \mathrm{USC}(\overline{\Omega})$ and $w \in \mathrm{LSC}(\overline{\Omega})$, which are respectively $\mathcal{F}$-subharmonic and $\mathcal{F}$-superharmonic on Ω. By duality, this is equivalent to showing

$$u + v \leq 0 \text{ on } \partial\Omega \implies u + v \leq 0 \text{ on } \Omega \tag{1.43}$$

for all $u, v \in \mathrm{USC}(\overline{\Omega})$, which are respectively $\mathcal{F}$-subharmonic and $\widetilde{\mathcal{F}}$-subharmonic on Ω. Our method of proof for $\mathcal{M}$-monotone subequations $\mathcal{F}$ makes use of this second formulation.

Here is a guide to the method. There are four steps.

Step 1: Jet addition. We have the following elementary but important fact concerning constraint sets, monotonicity, and duality:

$$\mathcal{F} + \mathcal{M} \subset \mathcal{F} \iff \mathcal{F} + \widetilde{\mathcal{F}} \subset \widetilde{\mathcal{M}}.$$

So the monotonicity condition on the left is equivalent to the condition on the right, which is perfect for comparison, as one sees from (1.43).

Showing that this infinitesimal statement passes to a potential-theoretic statement is the hard analysis step in the method.

Step 2: Subharmonic addition. We prove the following potential-theoretic result.

Theorem 7.4 (Subharmonic addition, monotonicity, and duality). *Suppose that $\mathcal{M} \subset \mathcal{J}^2$ is a monotonicity cone subequation and that $\mathcal{F} \subset \mathcal{J}^2$ is an $\mathcal{M}$-monotone* *subequation constraint set. Then for every open set $X \subset \mathbb{R}^n$, one has*

$$\mathcal{F}(X) + \widetilde{\mathcal{F}}(X) \subset \widetilde{\mathcal{M}}(X)$$

(where $\mathcal{F}(X)$ is the set of $u \in \mathrm{USC}(X)$ which are $\mathcal{F}$-subharmonic on X).

Step 3: Reduce comparison to the *zero maximum principle* (ZMP) for $\widetilde{\mathcal{M}}$. Armed with Theorem 7.4, it is clear from (1.43) that comparison for $\mathcal{F}$ on Ω will hold if we can prove the (ZMP) for $\widetilde{\mathcal{M}}$ on a bounded domain $\Omega \subset \mathbb{R}^n$, that is,

$$z \le 0 \text{ on } \partial\Omega \quad \Rightarrow \quad z \le 0 \text{ on } \Omega \tag{1.44}$$

for all $z \in \mathrm{USC}(\overline{\Omega})$ which are $\widetilde{\mathcal{M}}$-subharmonic on Ω.

Step 4: Prove the (ZMP) for $\widetilde{\mathcal{M}}$. A key concept in the proof is the following.

Definition 1.22. Suppose that $\mathcal{M}$ is a convex cone subequation. Given a domain $\Omega \subset\subset \mathbb{R}^n$, we say that $\mathcal{M}$ *admits a strict approximator on Ω* if there exists ψ with

$$\psi \in C(\overline{\Omega}) \cap C^2(\Omega) \quad \text{and} \quad J_x^2 \psi \in \mathrm{Int}\,\mathcal{M} \text{ for each } x \in \Omega. \tag{1.45}$$

This important notion gives a sufficient condition for proving the (ZMP) for $\widetilde{\mathcal{M}}$ and hence comparison for $\mathcal{F}$.

Theorem 6.2 (Zero maximum principle). *Suppose that $\mathcal{M}$ is a convex cone subequation that admits a strict approximator on Ω. Then the (ZMP) holds for $\widetilde{\mathcal{M}}$ on $\overline{\Omega}$.*

Putting these four steps together gives the following theorem.

Theorem 7.5 (General comparison theorem). *Suppose that $\mathcal{M} \subset \mathcal{J}^2$ is a monotonicity cone subequation and that $\mathcal{F} \subset \mathcal{J}^2$ is an $\mathcal{M}$-monotone subequation constraint set. Suppose that $\mathcal{M}$ is a convex cone subequation that admits a strict approximator on Ω. Then comparison holds for $\mathcal{F}$ on $\overline{\Omega}$. That is, given $u, v \in \mathrm{USC}(\overline{\Omega})$, where u is $\mathcal{F}$-subharmonic on Ω and v is $\widetilde{\mathcal{F}}$-subharmonic on Ω, then*

$$u + v \le 0 \text{ on } \partial\Omega \quad \Rightarrow \quad u + v \le 0 \text{ on } \Omega.$$

The conclusion here can be restated as follows. Given $u \in \mathrm{USC}(\overline{\Omega})$ and $w \in \mathrm{LSC}(\overline{\Omega})$, where u is $\mathcal{F}$-subharmonic on Ω and w is $\mathcal{F}$-superharmonic on Ω, then

$$u \le w \text{ on } \partial\Omega \quad \Rightarrow \quad u \le w \text{ on } \Omega.$$

To see this last statement we only need to know that w is $\mathcal{F}$-superharmonic on Ω if and only if $v \equiv -w$ is $\widetilde{\mathcal{F}}$-subharmonic on Ω.

Remark 1.23 (Affine jet equivalence). Theorem 7.5 generalizes from constant-coefficient subequations $\mathcal{F}$ to subequations $\mathcal{F}$ which are *locally affinely jet equivalent* to a constant-coefficient subequation. For any such equation, *local and hence global weak comparison holds*. Furthermore, if there exists a strict approximator (a classical strict $\mathcal{M}$-subharmonic) on Ω, then *comparison holds* on Ω. The reader should see [49] for all the details.

We point out that this concept is very useful in geometry. Any invariant polynomial equation, such as the Monge–Ampère equation or the elementary symmetric function equations, on a Riemannian manifold are always locally jet equivalent to constant-coefficient equations in local coordinates. A similar statement holds on almost complex (and therefore complex) manifolds. The reader should consult [54, Proposition 4.5].

Now, the utility of the general comparison theorem (Theorem 7.5) is greatly enhanced by a detailed study of monotonicity cone subequations, which we present in Chapter 5. There is a three-parameter *fundamental family* of monotonicity cone subequations $\mathcal{M}(\gamma, \mathcal{D}, R)$, where $\mathcal{D} \subset \mathbb{R}^n$ is a directional cone, $\gamma \in [0, \infty)$, and $R \in (0, \infty]$. In the fundamental family theorem (Theorem 5.10), it is shown that

> every monotonicity cone subequation $\mathcal{M}$ contains one of these subequations $\mathcal{M}(\gamma, \mathcal{D}, R)$.

We note that

$$\mathcal{M}(\gamma) := \left\{ (r, p, A) \in \mathcal{J}^2 : r \le -\gamma|p| \right\}, \quad \mathcal{M}(R) := \left\{ (r, p, A) \in \mathcal{J}^2 : A \ge \tfrac{|p|}{R} I \right\},$$

and

$$\mathcal{M}(\gamma, \mathcal{D}, R) := \mathcal{M}(\gamma) \cap \mathcal{M}(\mathcal{D}) \cap \mathcal{M}(R),$$

where $\mathcal{M}(\mathcal{D}) := \mathbb{R} \times \mathcal{D} \times \mathcal{S}(n)$.

The fundamental nature of this family of monotonicity cones, together with the general comparison principle of Theorem 7.5, leads to a main comparison result Theorem 7.6 (the fundamental family comparison theorem), which depends on the cone $\mathcal{M}(\gamma, \mathcal{D}, R)$. For some of these cones, comparison holds on all bounded domains. For the others, comparison holds only on domains $\Omega \subset \mathbb{R}^n$ which are subsets of a translation of the truncated cone $\mathcal{D} \cap B_R(0)$. This is a semilocal comparison with explicit parameters. Note that by the fundamental families theorem (Theorem 5.10), local comparison always holds (see Theorem 7.8).

Concerning the applicability of the fundamental comparison result of Theorem 7.6, it is worth mentioning that larger monotonicity cones $\mathcal{M}$ for a given subequation $\mathcal{F}$ give a better chance of proving comparison (one more likely to be able to construct a strict approximator) but smaller monotonicity cones $\mathcal{M}$ will apply to larger families of subequations. In particular, if one would like to know whether comparison holds on arbitrary bounded domains, one should search for the largest possible $\mathcal{M}$, which is perhaps not in the list of the fundamental family. For example, in Theorems 8.3 and 8.5 we present enlargements of the cones with R finite for which comparison holds on all bounded domains.

On the other hand, the search for sufficient monotonicity to have comparison on arbitrary bounded domains may be futile. In particular, for $\mathcal{F} := \mathcal{M}(R)$, which is its own maximal monotonicity cone, it is shown in Proposition 6.5 that the (ZMP) fails for $\widetilde{\mathcal{M}}(R)$ on large balls, and hence comparison also fails for $\mathcal{F} = \mathcal{M}(R)$ on large balls. This failure of comparison on large balls is extended to interesting subequations $\mathcal{F}$ with maximal monotonicity cone equal to $\mathcal{M}(R)$ in Proposition 9.2. The situation can be even worse.

Remark 1.24 (Failure of local comparison with insufficient monotonicity). In Theorem 9.8 we show that comparison can fail on arbitrarily small balls (even if both (P) and (N) hold) if there is insufficient monotonicity. In the examples the maximal monotonicity cone $\mathcal{M}_{\mathcal{F}}$ has empty interior, hence no strict approximators on any ball, no matter how small. Moreover, if $\mathcal{M}_{\mathcal{F}}$ has empty interior, then its dual is not a subequation.

Concerning step 3 of our method (in which comparison reduces to the validity of the (ZMP) for the dual $\widetilde{\mathcal{M}}$ of the monotonicity cone), the following observation is of interest.

Remark 1.25 (Strong comparison from the strong (ZMP)). The monotonicity and duality method can be used to prove a strong comparison principle which, by the subharmonic addition theorem, reduces to proving a strong (ZMP) for $\widetilde{\mathcal{M}}$ on $\overline{\Omega}$; that is,

$$z \le 0 \text{ on } \partial\Omega \;\Rightarrow\; z \equiv 0 \text{ or } z < 0 \text{ on } \Omega \qquad (1.46)$$

for all $z \in \mathrm{USC}(\overline{\Omega})$ which are $\widetilde{\mathcal{M}}$-subharmonic on Ω. This method was used in [55] to prove strong comparison for pure second-order subequations. We will not attempt to extend this to the general constant-coefficient case in this book. There is, of course, a rich literature on the strong maximum principle for nonlinear operators, including the important work of Bardi–Da Lio initiated in [4], along with the recent papers by Birindelli–Galise–Ishii [7], Vitolo [90], and Goffi–Pediconi [42].

A few additional potential-theoretic ingredients are worth mentioning. First, an elaboration on the potential theory underlying Example 1.8.

Remark 1.26 (Canonical operators, duality, intersections, and unions). For families $\{\mathcal{F}_\sigma\}_{\sigma \in \Sigma}$ of subequations with a common monotonicity cone subequation $\mathcal{M}$, by using unions, intersections, and duality, four interesting $\mathcal{M}$-monotone subequations are constructed, together with their canonical operators (see Theorem 11.23).

Next, an elaboration on the gradient-free case.

Remark 1.27 (Subaffine-plus functions). Subaffine-plus theory concerns the potential theory of the gradient-free subequation

$$\widetilde{Q} := \big\{ (r, A) \in \mathbb{R} \times \mathcal{S}(n) : r \leq 0 \text{ or } A \in \mathcal{P} \big\}, \qquad (1.47)$$

where $\widetilde{\mathcal{P}}$ is the pure second-order *subaffine* subequation. This $\widetilde{Q}$ is the dual of the fundamental gradient-free monotonicity cone $\mathcal{M} = \mathcal{Q} := \mathcal{N} \times \mathcal{P}$ and this potential theory is developed in detail (see Theorems 10.7 and 10.8). In particular, we extend the elegant method of using subaffine functions (the $\widetilde{\mathcal{P}}$-subharmonics) to prove that "comparison always holds" for pure second-order subequations. Subaffine-plus functions (the $\widetilde{Q}$-subharmonics) are used to prove that "comparison always holds" for the larger family of gradient-free subequations.

1.8   LIMITATIONS OF THE METHOD AND COMPARISON WITH THE LITERATURE

The monotonicity and duality method presented here applies to a vast array of constant-coefficient potential theories and operators, but not all of them. There are many interesting and important examples with insufficient monotonicity to be treated by our method. For example, quasi-linear operators such as the *minimal surface operator*

$$F(p, A) := \operatorname{tr}(A) - \frac{\langle Ap, p \rangle}{1 + |p|^2},$$

the *q-Laplacian* (with $1 < q < 2$ or $2 < q < \infty$)

$$F(p, A) := |p|^{q-2} \operatorname{tr}(A) + (q - 2)|p|^{q-4} \langle Ap, p \rangle,$$

and the *infinite Laplacian*

$$F(p, A) := \langle Ap, p \rangle,$$

do not have monotonicity cones $\mathcal{M}$ with interior, which we require. Such examples (and others) have been treated by Barles–Busca [6] by using an ingenious transformation of the dependent variable, which still cries out for a potential-theoretic analogue. On the other hand, for these reduced operators F (no explicit dependence on the jet variable $r \in \mathbb{R}$), Barles and Busca require structural assumptions on F such as their condition (F2): F *is strictly elliptic in the gradient direction*. This condition is *not* satisfied by an operator such as

$$F(p, A) := d(p) \det A,$$

which is Example 1.16 with $h(r) \equiv 1$ and $\mathfrak{g}(A) = \det A$. Such examples can be treated by our method.

Next we discuss a prototype operator which, surprisingly, creates difficulty for any method. The operator looks particularly attractive for comparison since it is strictly decreasing in the solution variable $r \in \mathbb{R}$ and is increasing in the Hessian variable A when restricted to $\mathcal{P} \subset \mathcal{S}(n)$. Namely, consider the seemingly innocuous operator

$$F(r, A) = \det A - r \qquad (1.48)$$

(which is further discussed in Remark 12.7). The operator F is gradient-free and proper elliptic on $\mathbb{R} \times \mathcal{P}$; that is, it is $\mathcal{Q} = \mathcal{N} \times \mathcal{P}$-monotone on $\mathcal{F} := \mathbb{R} \times \mathcal{P}$. However, the potential theory equation $\partial \mathcal{F}$ is not contained in the zero locus $\{(r, A) \in \mathcal{F} : F(r, A) = 0\}$, that is, F and $\mathcal{F}$ are not compatible. This cannot be remedied by another choice of $\mathcal{F}$, creating a major obstacle to the study of this operator. This incompatibility means that $\mathcal{F}$-superharmonics will not correspond to $\mathcal{F}$-admissible supersolutions to the equation $F(u, Du, D^2u) = 0$. In order to formulate a notion of admissible supersolution, one could make use of the *generalized equation* approach initiated in [61] for pure second-order equations in which one looks for a second constraint set $\mathcal{G}$ (different from $\mathcal{F}$) such that

$$\mathcal{F} \cap (-\widetilde{\mathcal{G}}) = \{(r, A) \in \mathcal{F} : F(r, A) = 0\}.$$

The admissible supersolutions are those $w \in \mathrm{LSC}(\Omega)$ which are $-\widetilde{\mathcal{G}}$-subharmonic. We will not pursue this program here.

In addition to the paper [6] discussed above, earlier pioneering work in the constant-coefficient case was done by Jensen [69]. The equations treated by him are all unconstrained in our language, where the monotonicity properties (P) and (N) do *not* require restricting the domain F to a constraint set $\mathcal{F}$. In Sections 12.2 and 12.3, we recover Jensen's results in this unconstrained setting (see Remark 12.10). Of course, we also treat many constrained cases in this monograph, which is an important motivation for us.

Concerning the constrained case and our notion of compatible pairs $(F, \mathcal{F})$, we should mention that the special case of Monge–Ampère-type equations with the convexity constraint $\mathcal{P}$ is given in Ishii–Lions [68], together with a notion of admissible supersolutions in our language. A similar admissibility notion was also given by Trudinger [87] for prescribed curvature equations, and later by Trudinger–Wang for the so-called Hessian equations in a series of papers beginning with [88]. As noted previously, another motivation of ours is to treat constrained cases in a robust and general way. The potential-theoretic approach initiated in [46] was influenced by the important paper of Krylov [77] on the general notion of ellipticity, who championed the idea of freeing a given differential operator F from its particular form by looking instead at the constraint that is imposed on the 2-jets of subsolutions to the equation.

Finally, we wish to comment on our choice to focus on the constant-coefficient case. The most basic reason is that in this situation, monotonicity and duality alone suffice to produce comparison for compatible pairs $(F, \mathcal{F})$. Much more can be said when dependence on spatial coordinates is added into the pair, or one works on manifolds, but additional conditions must be imposed in order to prove the comparison principle. We briefly review three situations in which the monotonicity–duality method has been extended past the constant-coefficient setting.

On open sets X in $\mathbb{R}^n$, one can add x-dependence in two ways. The first is by considering *inhomogeneous equations* associated to a constant-coefficient compatible pair $(F, \mathcal{F})$, that is, equations of the form

$$F(J^2 u) = \psi(x), \quad x \in X. \tag{1.49}$$

Here, one should consider $\psi \in C(X)$ taking values in the range $F(\mathcal{F})$ of the operator $F \in C(\mathcal{F})$.

As shown in [60], if $(F, \mathcal{F})$ is a compatible $\mathcal{M}$-monotone pair for some monotonicity cone subequation $\mathcal{M}$ which admits a classical strict subharmonic function, then the *tameness* condition of F on $\mathcal{F}$,

for each $s, \lambda > 0$ there exists $c(s, \lambda) > 0$ such that

$$F(J + (-r, 0, P)) - F(J) \geq c(s, \lambda) \quad \forall J \in \mathcal{F}, \ \forall r \geq s, \ \forall P \geq \lambda I, \tag{1.50}$$

ensures that the comparison principle holds for $\mathcal{F}$-admissible subsolutions and supersolutions of (1.49). The tameness property is crucial to maintain the $\mathcal{F}$-subharmonicity of sup-convolutions of $\mathcal{F}$-subharmonic functions.

This comparison result is part of the content of [60, Theorem 2.7], which also treats the existence of a unique solution to the Dirichlet problem on bounded domains Ω which are boundary pseudoconvex in a suitable strict sense. See also [60, Theorem 2.7′] for the extension to operators F which are *tamable*. Moreover, [60, Theorem 2.7] extends to manifolds X for pairs $(F, \mathcal{F})$ which are *locally jet-equivalent* to a constant-coefficient pair as above (see [60, Theorem 2.11]).

On open sets X in $\mathbb{R}^n$, a more general way to add in x-dependence is to consider general operators $F \in C(\mathcal{G})$ where either $\mathcal{G} = J^2(X)$ or $\mathcal{G} \subsetneq J^2(X)$ is a subequation constraint set. The candidate for a compatible subequation $\mathcal{F} \subset J^2(X)$ for F is defined fiberwise by the *correspondence relation*

$$\mathcal{F}_x := \{ J \in \mathcal{G}_x : F(x, J) \geq 0 \}, \quad x \in X. \tag{1.51}$$

In this setting, *fiberegularity* is crucial. This condition means that the fiber map Θ, with values $\Theta(x) := \mathcal{F}_x$ as defined in (1.51), is continuous from $X \subset \mathbb{R}^n$ into the closed subsets of $\mathcal{J}^2$ (equipped with the Hausdorff distance). For $\mathcal{M}$-monotone subequations, fiberegularity has a more useful equivalent formulation:

there exists $J_0 \in \operatorname{Int} \mathcal{M}$ such that, for each fixed $\Omega \subset\subset X$ and $\eta > 0$, there exists $\delta = \delta(\eta, \Omega)$ such that

$$x, y \in \Omega, \ |x - y| < \delta \ \Rightarrow \ \Theta(x) + \eta J_0 \subset \Theta(y). \tag{1.52}$$

One can show that if there exists such a jet J_0, then any element of $\operatorname{Int} \mathcal{M}$ will do, where one can simply take $J_0 = (-1, I) \in \mathbb{R} \times \mathcal{S}(n)$ or $J_0 = I \in \mathcal{S}(n)$ in the gradient-free and pure second-order cases, respectively.

Fiberegularity ensures that if u is $\mathcal{F}$-subharmonic on a bounded domain, then there are small C^2-strictly $\mathcal{F}$-subharmonic perturbations of all small translates of u which belong to $\mathcal{F}(\Omega_\delta)$, where $\Omega_\delta := \{x \in \Omega : d(x, \partial\Omega) > \delta\}$. This property is crucial to maintain $\mathcal{F}$-subharmonicity of sup-convolution approximations of $\mathcal{F}$-subharmonics. Hence, in the general setting, fiberegularity plays the same role as tameness does for inhomogeneous equations (as noted after (1.50)).

Fiberegularity, together with monotonicity and duality, has been shown to be sufficient for comparison in the pure second-order, gradient-free, and $\mathcal{M}$-monotone cases in [23–25]. See also the recent paper of Brustad [12]. In particular, this leads to the proof of the comparison principle [24] for the special Lagrangian potential equation with nonconstant phases (as introduced in Example 1.9) provided that the phase function does not take on a *special phase value*. This result is sharp, as shown in Brustad [13]. As noted in the preface, this potential-theoretic approach has been used to prove the comparison principle for PDEs that do not satisfy a standard structural condition from conventional viscosity theory. The following is a simple pure second-order example from [23]. Replacing det with a different *Gårding–Dirichlet polynomial* $\mathfrak{g}$ (as defined in Section 1.4) yields a huge family of such pure second-order examples that can be further generalized by taking $F(r, p, A) = h(r, p)\mathfrak{g}(A)$ for suitable $h \in C(\mathbb{R} \times \mathbb{R}^n)$.

Example 1.28 (Perturbed Monge–Ampère). With fixed $M \in C(\Omega, \mathcal{S}(n))$ and $f \in C(\Omega)$ nonnegative, consider

$$\det(D^2 u + M(x)) = f(x), \quad x \in \Omega \subset\subset \mathbb{R}^n. \tag{1.53}$$

This is an important test example of Krylov [76, Example 8.2.4] for probabilistic and analytic methods. It fails to satisfy the standard viscosity structural conditions for comparison as given in Crandall–Ishii–Lions [30, condition (3.14)] unless M is the square of a Lipschitz continuous matrix-valued function. In [23], comparison is proved for general continuous M (along with the existence of a unique continuous solution of the Dirichlet problem on strictly convex domains). The potential-theoretic proof makes use of the compatible subequation whose fibers are defined by

$$\mathcal{F}_x := \big\{ A \in \mathcal{S}(n) : A + M(x) \geq 0 \text{ and } F(x, A) := \det(A + M(x)) - f(x) \geq 0 \big\},$$

and give a fiberegular $\mathcal{F}$. The strict convexity is the boundary geometry required by $\mathcal{F}$ as described after Remark 1.6.

Finally, constant-coefficient subequations on Euclidean space generate a rich and interesting class of subequations on manifolds X, as developed in [49]. These subequations on X are those which are *locally jet-equivalent to a constant-coefficient subequation*. Any Riemannian G-subequation on a Riemannian manifold X with topological G-structure is such a subequation. For simple examples, let $\mathfrak{p} \colon \mathcal{S}(n) \to \mathbb{R}$ be a continuous function which is invariant under the action of $\mathrm{O}(n)$ (such as the determinant or the trace). Applying $\mathfrak{p}$ to the Riemannian Hessian gives an operator (real Monge–Ampère or Laplace–Beltrami) on X, which has the jet-equivalence property above. These notions are discussed in [49, pp. 398–402], along with much more.

1.9 REFLECTIONS ON "POTENTIAL THEORY VERSUS OPERATOR THEORY"

The work of Harvey–Lawson on generalized potential theories began with the realization that every calibrated manifold had an underlying potential theory, which generalized much of the basic pluripotential theory associated to the special case of Kähler manifolds. It soon became clear (but surprising) that the set of p-planes distinguished by the calibration, which gives rise to the pluripotential theory, could be replaced by any closed subset of the p-Grassmannian, yielding an interesting geometric potential theory. Over time it was found that these potential theories have far-reaching generalizations, which give new dimensions to interesting areas of geometry. The third and final step was to bypass the subset of the Grassmannian and go directly to a constraint set on the 2-jets of a function. This investigation has led to a constellation of potential theories as discussed in the survey paper [65, Section 1.1].

Each such potential theory has a basic geometry which gives rise to a large number of differential operators that can be used in the analysis. In fact, in important situations (for example, in $\mathrm{G}(2)$ and $\mathrm{Spin}(7)$ geometries) there are no historical "polynomial" operators, but one does have the canonical operators and the signed distance operators, which are determined by the geometry of the constraint set. A good example of this is the introduction [46] of the truncated Laplacian for the geometric p-convexity constraint set. This lack of known polynomial operators was discussed in Section 1.4.

On the other hand, if one starts with a nonlinear proper elliptic operator F, one has an associated potential theory coming from the set $\mathcal{F}$ where $F \geq 0$. This potential theory is useful because it gives a different way of thinking; many of the *results* and *techniques* of classical potential theory carry over to these cases, and have led to new results.

One example is the analogue of the Levi problem in several complex variables, established in the new p-geometry, where all p-planes are considered distinguished

[52]. Namely, local p-convexity implies global p-convexity. See the review in [65, Section 1.2] for additional potential-theoretic results suggested by several complex variable theory.

A really important example is given by the sharp L^∞-estimates for the complex Monge–Ampère equation, for which deep results in pluripotential theory were used. The first major advance after Yau's fundamental paper [91] was given by Kolodziej [73]. His work had profound significance and was used for many future developments, and it relied heavily on pluripotential theory! So also did future generalizations of Kolodziej's work by Demailly–Pali [33] and Eyssidieux–Guedj–Zeriahi [38]. This work was crucial in completing the existence of Calabi–Yau metrics in the positive case, a long research program recently finished in [19–21].[4] So in this area, at the center stage of modern research, pluripotential theory was crucial in solving an important differential equation.

Two important advantages of a potential-theoretic approach to nonlinear PDEs are worth repeating. First, as mentioned at the beginning of the introduction, many operators F correspond to the same subequation $\mathcal{F}$ and passing directly to the potential theory "frees" one from any particular form of the differential operator. In the viscosity literature, there is often the need to find an ad hoc reformulation of the operator as a first step, which becomes unnecessary if one passes to the potential theory. This is a major point in the work of Krylov [77]. Second, as mentioned after Remark 1.6, the potential theory approach correctly identifies the needed boundary geometry for existence by Perron's method for any operator F that corresponds to a subequation $\mathcal{F}$ defining the potential theory [46, 49].

The potential theory viewpoint makes PDE concepts purely geometric. It is important in relating different subequations (by containment, intersection, etc.), which leads to understanding things in a new way. It is also important for generalizations of the notion of equation; for example, see [61].

Another accomplishment of the potential theory viewpoint is the transfer of results for important equations to manifolds with Riemannian, symplectic, complex, or almost complex structure. If $\mathcal{F}$ is a subequation, which is invariant under the natural action of $O(n)$, then $\mathcal{F}$ naturally determines a subequation on every Riemannian manifold. Similarly, if $\mathcal{F}$ is $U(n)$-invariant, one gets a subequation on every complex (or, even, almost complex) hermitian manifold. If $\mathcal{F}$ is $G(2)$-invariant, it defines a subequation on any (almost) $G(2)$ manifold, and so on. This, together with a notion of *affine jet equivalence*, gives wide-ranging results for the Dirichlet problem on domains in such manifolds [49, 50, 60], and more.

There are many recent papers which have picked up the major theme of this book. We mention now several illustrations.

---

[4]In fact, just recently, in an important paper [43] by Guo–Phong–Tong, these estimates were established by pure PDE methods. The breakthrough enabling [43] was a fundamental new idea of Chen–Cheng [18].

First, the interplay between potential theory and geometry via interesting differential operators on Riemannian manifolds has taken inspiration from [49]. In Mari–Pessoa [81] and Araújo–Mari–Pessoa [2], the potential theory of the infinite Laplacian and the Eikonel operator are used for characterizing various *maximum principles at infinity* and detecting the *forward completeness* of Finsler manifolds, respectively. There is also Goffi–Pediconi [42] on strong maximum principles for operators modeled on Pucci extremal operators, infinite Laplacians, and mean curvature operators.

Second, on hermitian manifolds which are not Kähler, potential-theoretic techniques have been used to construct *Gauduchon metrics* with prescribed volume form in Székelyhidi–Tosatti–Weinkove [86].

Third, on manifolds with an almost complex structure, there are many results that have taken inspiration from [49, 50]. For example, in [54] the Dirichlet problem was solved as well as the closely related *Pali conjecture* stating that distributional plurisubharmonics and classical plurisubharmonics are equivalent. Consequently, the solution to the *obstacle problem* in [49] applies to almost complex manifolds and yields smooth approximations from above of plurisubharmonic functions in [63]. See also [64].

Fourth, following the work of [46] on Euclidean spaces, the interplay between potential theory and (viscosity) operator theory is the major theme of the papers [23–25] treating in wide generality the comparison principle and Perron's method in the variable coefficient fiberegular setting. In addition, this interplay is seen in the solution of the plateau problem for convex hypersurfaces of constant Gaussian curvature in Clark–Smith [26] and in many works dedicated to various forms of the maximum principle as in Amendola–Galise–Vitolo [1] and Birindelli–Galise–Ishii [7, 9], as well as the study of entire subsolutions in Capuzzo Dolcetta–Leoni–Vitolo [16].

Fifth, the many geometric operators with an interesting potential theory developed in [46, 49] (and in subsequent papers) has provided a notable stimulus to many recent investigations. The special Lagrangian potential operator was introduced in calibrated geometry [44], as discussed above in Example 1.9. Providing the proof of comparison for the special Lagrangian potential equation (with constant admissible phases) was one of the major motivations for [46]. This operator is the subject of a long list of papers beginning with Chen–Warren–Yuan [17] and continuing with Nadirashvili–Vlăduţ [82], Rubinstein–Solomon [84], the work of the authors [24, 62], and many others including the very recent paper by Brustad [13]. Also recently, in this direction, has been the fundamental work of Collins–Yau in a series of three papers starting with [27]. The truncated Laplacian operators were introduced in the context of the geometrically p-plurisubharmonic Dirichlet problem [46], as discussed in Example 1.13. These operators have been the object of much recent work including Vitolo [90] and Birindelli–Galise–Ishii [7–9].

Of course, techniques from viscosity theory, in particular the theorem on sums, play a big role in the analysis in [49] and elsewhere. A brief review of some milestones in the history of viscosity methods follows.

The notion of viscosity solutions originates for first-order equations in Crandall–Lions [31], and has its roots in Kruzkov's theory of entropy solutions for conservation laws [74]. The basic idea to put derivatives on test functions by way of the maximum principle originated in the work of Evans [34, 35] on the weak limits in fully nonlinear PDEs by means of Minty's method. One of the major achievements of the viscosity theory has proven to be the treatment of difficult problems by appropriate limiting procedures. There are many illustrations of this achievement when making use of *homogenization* and/or *vanishing viscosity* parameters; for example, see Evans [36, 37]. The notion of viscosity solutions was extended to second-order equations by Lions [79, 80], who first gave a proof of uniqueness using stochastic control arguments when the operator F is convex or concave in (Du, D^2u).

A major breakthrough was the work of Jensen [69, 70] which frees the theory from its dependence on the convexity (or concavity) and makes use of regularizations by way of the sup and inf convolutions. This was refined in Jensen–Lions–Souganidis [71]. See also Remark 12.10, which discusses the relation between two of Jensen's comparison principles and what we obtain by our method.

The analytical underpinnings of the comparison principle were further reformulated making use of the technique of doubling variables and penalization and culminates in the theorem on sums of Crandall–Ishii [29], which in turn had its origins in Ishii [67], Ishii–Lions [68], and Crandall [28]. The theorem on sums was used in the treatment of comparison principles on manifolds in [49].

Another important breakthrough in viscosity theory was the introduction of the comparison principle as a tool for proving existence for the Dirichlet problem via Perron's method. This was first accomplished by Ishii in two landmark papers: first-order equations in [66] and second-order equations in [67]. What is often known as *Ishii's theorem* states that one has the existence of a unique solution provided that the comparison principle holds and provided that there is a good subsolution/supersolution pair. In practice, these two hypotheses may seem unrelated, but the potential theory approach connects the two.

There are at least two major achievements of the conventional viscosity theory which, at least for now, have not been replicated by the potential theory approach. The first is the ability to prove the comparison principle for important operators, such as the minimal surface operator, which lack sufficient monotonicity to use our method (see the discussion in Section 1.8). The second is that the use of limiting procedures via artificial viscosity and/or homogenization parameters has yet to be attempted in the potential-theoretic setting. These two questions provide impetus for further interplay.

We have mentioned the seminal contribution of Jensen [69] to comparison theory. A main result in this work has become known as *Jensen's lemma* (for example, see [30, Lemma A.3]). It is a measure-theoretic result for quasi-convex[5]

---

[5]Such functions are often referred to as *semiconvex* functions, although the term is a bit misleading. Our use of quasi-convex is consistent with the use of *quasi-plurisubharmonic* functions w in several complex variables.

functions which ensures a wealth of linear perturbations having local maxima in the neighborhood of a strict local maximum of w. It is interesting to note that this lemma is equivalent to a result of Slodkowski [85] (two years earlier) developed in the context of pluripotential theory and can be stated in terms of the measure of the set of *upper contact points* near a strict upper contact point. The equivalence of the Jensen and Slodkowski lemmas is given in [57, Theorem 3.6], which also provides another proof of the Slodkowski lemma using contact paraboloids in place of contact spheres. See also [83], which borrows heavily from [56], for additional reflections on this equivalence and the role of Alexandroff's maximum principle and the area formula of Federer, from which one can prove both lemmas. This equivalence should be more well known than it is and signals an important moment when possible synergy between potential theory and operator theory was missed. It is a major theme of this book to examine the possibly fruitful interplay between potential theory and operator theory.

Part II

The Potential Theory Approach

Chapter Two

Constant-Coefficient Constraint Sets and
Their Subharmonics

In this chapter we will discuss nonlinear potential theory. Two definitions are of fundamental importance: that of *subequation constraint sets* and that of their *subharmonics* (see Definitions 2.1 and 2.4). In all that follows, X will denote an open subset of $\mathbb{R}^n$, $\mathcal{S}(n)$ the space of symmetric $(n \times n)$-matrices with real entries (with its partial ordering given by the associated quadratic forms), and

$$\mathcal{J}^2 := \mathbb{R} \times \mathbb{R}^n \times \mathcal{S}(n) \tag{2.1}$$

will denote the space of *2-jets* with coordinates $J = (r, p, A)$. The spaces of upper-, lower-semicontinuous functions on X taking values in $[-\infty, +\infty)$, $(-\infty, +\infty]$ will be denoted by $\mathrm{USC}(X)$, $\mathrm{LSC}(X)$ respectively.

Definition 2.1. A *subequation (constraint set)* is a nonempty proper subset[1] $\mathcal{F} \subset \mathcal{J}^2$ which satisfies the *positivity condition*

$$(r, p, A) \in \mathcal{F} \text{ and } P \geq 0 \ \Rightarrow \ (r, p, A + P) \in \mathcal{F}, \tag{P}$$

the *negativity condition*

$$(r, p, A) \in \mathcal{F} \text{ and } s \leq 0 \ \Rightarrow \ (r + s, p, P) \in \mathcal{F}, \tag{N}$$

and the *topological condition*

$$\mathcal{F} = \overline{\mathrm{Int}\, \mathcal{F}}, \tag{T}$$

which implies that $\mathcal{F}$ is closed and has nonempty interior.

Denoting $\mathcal{P} := \{P \in \mathcal{S}(n) : P \geq 0\}$ and $\mathcal{N} := \{s \in \mathbb{R} : s \leq 0\}$, the monotonicity conditions are

$$\mathcal{F} + (\{0\} \times \{0\} \times \mathcal{P}) \subset \mathcal{F} \tag{P}$$

and

$$\mathcal{F} + (\mathcal{N} \times \{0\} \times \{0\}) \subset \mathcal{F}. \tag{N}$$

---

[1] Somewhat surprisingly, the nonempty and proper subset hypothesis is rarely needed in the proofs; for example, one always has (trivially) comparison if $\mathcal{F} = \emptyset$ or $\mathcal{F} = \mathcal{J}^2$.

Also, it is useful to note that for closed convex sets $\mathcal{F}$,

$$\text{the topological condition (T) holds} \quad \Leftrightarrow \quad \text{Int}\,\mathcal{F} \neq \emptyset. \tag{2.2}$$

Hence, a closed convex set $\mathcal{F} \subset \mathcal{J}^2$ is a subequation if and only if $\mathcal{F}$ satisfies (P) and (N) and has nonempty interior.

In addition to the monotonicity properties (P) and (N), a third monotonicity condition plays an important role in comparison and is introduced here. It depends on the choice of a suitable cone $\mathcal{D} \subseteq \mathbb{R}^n$.

Definition 2.2. A closed convex cone $\mathcal{D} \subseteq \mathbb{R}^n$ (possibly all of $\mathbb{R}^n$) with vertex at the origin, which satisfies the topological condition (T) (equivalently Int $\mathcal{D} \neq \emptyset$), will be called a *directional cone*. The *directionality condition* on a subequation $\mathcal{F} \subset \mathcal{J}$ is

$$(r, p, A) \in \mathcal{F} \text{ and } q \in \mathcal{D} \;\Rightarrow\; (r, p+q, A) \in \mathcal{F}, \tag{D}$$

or equivalently,

$$\mathcal{F} + (\{0\} \times \mathcal{D} \times \{0\}) \subset \mathcal{F}. \tag{D}$$

Remark 2.3. In previous works, the sets $\mathcal{F}$ described in Definition 2.1 have been called both a *Dirichlet set* and a *subequation*. Here we will often use the shortened form "subequation" in place of the longer phrase "subequation constraint set" introduced in Definition 2.1.

A function $u \in C^2(X)$ *satisfies the subequation constraint $\mathcal{F}$ if*
$$J_x^2 u := (u(x), Du(x), D^2u(x)) \in \mathcal{F} \quad \text{for each } x \in X, \tag{2.3}$$

and will be called *$\mathcal{F}$-subharmonic on X.* If

$$J_x^2 u \in \text{Int}\,\mathcal{F} \quad \text{for each } x \in X, \tag{2.4}$$

we will say that u is *strictly $\mathcal{F}$-subharmonic on X.*

For upper-semicontinuous functions $u \in \text{USC}(X)$, one can define the differential inclusion (2.3) in the *viscosity sense.*

Definition 2.4. Given a function $u \in \text{USC}(X)$,

(a) a function φ which is C^2 near $x_0 \in X$ is said to be a *(C^2-upper) test function for u at x_0* if
$$u - \varphi \leq 0 \text{ near } x_0 \quad \text{and} \quad u - \varphi = 0 \text{ at } x_0; \tag{2.5}$$

(b) the function u is said to be *$\mathcal{F}$-subharmonic at x_0* if $J_{x_0}^2 \varphi \in \mathcal{F}$ for all upper C^2-test functions φ for u at x_0;

(c) the function u is said to be *$\mathcal{F}$-subharmonic on X* if u is $\mathcal{F}$-subharmonic at each $x_0 \in X$.

The space of all $\mathcal{F}$-subharmonic functions on X will be denoted by $\mathcal{F}(X)$.

A pair of remarks concerning $\mathcal{F}$-subharmonicity are in order.

Remark 2.5. There are several equivalent ways of defining u to be $\mathcal{F}$-subharmonic at $x_0 \in X$, which can all be formulated as

$$J = (r, p, A) \in \mathcal{F} \quad \text{for all "upper test jets" for } u \text{ at } x_0. \qquad (2.6)$$

To complete definition (2.6), there are several natural choices for defining the concept "$J = (r, p, A)$ is an upper test jet for u at x_0," which all yield the same notion of u being $\mathcal{F}$-subharmonic at x_0. This is, of course, well known to specialists. For the convenience of the reader, we present four equivalent reformulations in Lemma C.1. The statement includes nomenclature for each formulation that we believe to be useful. (Also, somewhat surprisingly, all of the equivalences of Lemma C.1 are valid for any closed set $\mathcal{F}$ without mention of properties (P) or (N).) Being $\mathcal{F}$-subharmonic at x_0 is the differential inclusion

$$J_*(x_0, u) \subset \mathcal{F} \quad \text{for one of the four choices of upper test jets in Lemma C.1.} \qquad (2.7)$$

It is important to note that the four sets of upper test jets for u at x_0 are nested in the sense that (see (C.2))

$$J_1(x_0, u) \subset J_2(x_0, u) \subset J_3(x_0, u) \subset J_4(x_0, u). \qquad (2.8)$$

Using the smallest set $J_1(x_0, u)$ of *strict quadratic test jets* gives the best choice for showing that u is subharmonic by a contradiction argument. This is done in the bad test jet lemma (Lemma 2.8), which concludes that there is a bad test jet $J \in J_1(x_0, u)$. The set $J_3(x_0, u)$ of C^2-*test jets* is the set of upper test functions used in our Definition 2.4(b) above. This definition is local, but can be reformulated as a form of the comparison principle, which always holds globally. See Lemma 3.14 on *definitional comparison*. Finally, the largest set $J_4(x_0, u)$ of *little-o quadratic test jets* yields one of the standard definitions in terms of the second-order superdifferential since $J_4(x_0, u) = J^{2,+}_{x_0} u$.

Remark 2.6. The pointwise notion of Definition 2.4(b) that u is $\mathcal{F}$-subharmonic at x_0 is not without its "pathology." For example, the function $u(x) = |x|$ is $\mathcal{F}$-subharmonic at $x_0 = 0$ for every subequation $\mathcal{F}$ since there are no upper test functions φ for u at $x_0 = 0$. However, the concept of u being $\mathcal{F}$-subharmonic on an open set X mitigates this pathology in the following way. For a fixed $u \in \mathrm{USC}(X)$, consider the set of *upper contact points*

$$\mathrm{UCP}(X) := \{x_0 \in X : J_*(x_0, u) \neq \emptyset\}, \qquad (2.9)$$

where $J_*(x_0, u)$ is any one of set of test jets in (2.8), as defined by (J1)–(J4) in Lemma C.1. One can show that $u \equiv -\infty$ on the open set $X \setminus \overline{\mathrm{UCP}(X)}$ by using [57, proof of Lemma 6.1]. In particular, if $u(x_0) > -\infty$ (u is finite at x_0), then x_0 is the limit of a sequence $\{x_k\}_{k \in \mathbb{N}} \subset \mathrm{UCP}(X)$.[2]

---

[2]This fact strengthens [72, Proposition 2.5(2)], where only USC-functions which are everywhere finite are considered.

The natural notion of u being $\mathcal{F}$-*harmonic on* X will be recorded in Definition 3.6, in terms of *Dirichlet duality* (along with the notion of u being $\mathcal{F}$-*superharmonic on* X).

A pair of elementary examples of $\mathcal{F}$-subharmonic functions are worth mentioning to help fix the idea.

Examples 2.7 (Convex functions and classical Laplacian subharmonics). For the *convexity subequation* $\mathcal{F} = \mathbb{R} \times \mathbb{R}^n \times \mathcal{P}$, one has (see [46, Proposition 4.5])

$$u \in \mathcal{F}(X) \iff u \text{ is convex or } u \equiv -\infty \text{ on connected components of } X. \quad (2.10)$$

For $\mathcal{F} = \mathbb{R} \times \mathbb{R}^n \times \mathcal{F}_\Delta$ with $\mathcal{F}_\Delta := \{ A \in \mathcal{S}(n) : \operatorname{tr}(A) \geq 0 \}$, one can show that for $u \in \mathrm{USC}(X)$,

$$u \in \mathcal{F}(X) \iff u(x_0) \leq \frac{1}{|B_r(x_0)|} \int_{B_r(x_0)} u(x)\, dx \text{ for each } B_r(x_0) \subset\subset X, \quad (2.11)$$

which is the classical definition of a subharmonic function.

In both examples, there is a canonical[3] choice (but not the only choice) of a differential operator $F(u, Du, D^2 u)$ with

$$\mathcal{F} = \big\{ (r, p, A) \in \mathcal{J}^2 : F(r, p, A) \geq 0 \big\},$$

and hence $\mathcal{F}(X)$ is the space of viscosity subsolutions of $F(u, Du, D^2 u) = 0$. The first is the *minimum eigenvalue operator*

$$F(r, p, A) = \lambda_{\min}(A),$$

and the second is the *Laplacian* $F(r, p, A) = \operatorname{tr}(A)$.

As noted in Remark 2.5, if one takes the contrapositive of Definition 2.4(b) using the *strict quadratic test jet* formulation (J1) of Lemma C.1, one immediately obtains the following very useful tool for establishing $\mathcal{F}$-subharmonicity by providing the existence of a "bad test jet" at a point where $\mathcal{F}$-subharmonicity fails.

Lemma 2.8 (Bad test jet lemma). *Given* $u \in \mathrm{USC}(X)$ *and* $\mathcal{F} \neq \emptyset$, *if* u *is not* $\mathcal{F}$-*subharmonic at* x_0 *then there exist* $\varepsilon > 0$ *and* $J = (r, p, A) \notin \mathcal{F}$ *such that the quadratic function* Q_J *with 2-jet* J *at* x_0 *is a test function for* u *at* x_0 *in the following* ε-*strict sense:*

$$u(x) - Q_J(x) \leq -\varepsilon |x - x_0|^2 \quad \text{near } x_0, \text{ with equality at } x_0. \quad (2.12)$$

---

[3]See [60, Proposition 6.11] for the definition of the *canonical operator*, its construction, and many additional examples. See also Section 11.4 below.

The converse of Lemma 2.8 is one of the many equivalent definitions of $\mathcal{F}$-subharmonicity (using the strict quadratic test jets $J_1(x_0, u)$).

Remark 2.9 (Coherence and the positivity condition). A fundamental observation concerning Definition 2.4(b) of $u \in \mathrm{USC}(X)$ being $\mathcal{F}$-subharmonic at x_0 with $\mathcal{F} \neq \emptyset$ is the following *coherence property*: if u is twice differentiable at x_0 then

$$u \text{ is } \mathcal{F}\text{-subharmonic at } x_0 \iff J^2_{x_0} u \in \mathcal{F}. \tag{2.13}$$

The forward half of the equivalence ($\Rightarrow$) is an immediate consequence of the little-o formulation (C.3) of Lemma C.1, where the second-order Taylor expansion of u in x_0 with Peano remainder is an upper test function for u at x_0. The reverse half of the equivalence ($\Leftarrow$) is the first instance where the positivity condition (P) for $\mathcal{F}$ is required. Indeed, assume $J^2_{x_0} u \in \mathcal{F}$ and that φ is an upper test function satisfying (2.5). By elementary calculus, one has $\varphi(x_0) = u(x_0)$, $D\varphi(x_0) = Du(x_0)$, and $D^2 u(x_0) - D^2\varphi(x_0) = -P$ with $P \geq 0$, and hence

$$J^2_{x_0}\varphi = J^2_{x_0} u + P \quad \text{with } P \geq 0,$$

which yields $J^2_{x_0}\varphi \in \mathcal{F}$ if and only if property (P) holds for $\mathcal{F}$.

Another result which makes use of property (P) is the following useful lemma, which has been given in a more general form on manifolds in [57, Theorem 4.1]. For the convenience of the reader we will give a sketch of the proof. We recall that a function u is *locally quasi-convex (locally semiconvex) on an open set* $X \subset \mathbb{R}^n$ if, for each $x \in X$ and for some $\lambda > 0$, the function $u(\cdot) + \frac{\lambda}{2}|\cdot|$ is convex on a neighborhood of x.

Lemma 2.10 (Almost everywhere theorem). *Suppose that $\mathcal{F} \subset \mathcal{J}^2$ is a subequation constraint set and that u is a locally quasi-convex function on the open set* $X \subset \mathbb{R}^n$. *Then u is twice differentiable almost everywhere on X by Alexandroff's theorem and*

$$J^2_x u \in \mathcal{F} \text{ for almost every } x \in X \implies u \text{ is } \mathcal{F}\text{-subharmonic on } X. \tag{2.14}$$

Proof. The proof is by contradiction. If u fails to be $\mathcal{F}$-subharmonic at $x_0 \in X$, then the bad test jet lemma (Lemma 2.8) yields a strict upper contact jet $J = (r, p, A) \notin \mathcal{F}$ for which (2.12) holds in a neighborhood of x_0. From this strict upper contact jet, the Jensen–Slodkowski lemma (see [57, Theorem 3.6]) applied to the subset $E \subset X$ of full measure on which u is twice differentiable at $x \in E$ and $J^2_x u \in \mathcal{F}$ yields a sequence $\{x_k\}_{k \in \mathbb{N}} \subset E$ such that, for $k \to +\infty$, one has

$$x_k \to x_0 \quad \text{and} \quad J^2_{x_k} u \to (r, p, A - P) \text{ with } P \geq 0.$$

Since $\mathcal{F}$ is closed, one has $(r, p, A - P) \in \mathcal{F}$ and then, by property (P), $J = (r, p, A) = (r, p, A - P + P) \in \mathcal{F}$, which contradicts the choice of J. $\square$

Remark 2.11. Of course, by Alexandroff's theorem and the coherence property of Remark 2.9, implication (2.14) is an equivalence.

We conclude this chapter with a few additional remarks on conditions (P), (N), and (T) of Definition 2.1 and condition (D) of Definition 2.2. Condition (P) corresponds to *degenerate ellipticity* of the differential inclusion (2.7) for each $x_0 \in X$ and, as shown above, proves to be crucial in showing one-half of the coherence property (2.13). Property (N), when combined with (P), corresponds to *properness* of (2.7) and is well known to play a role in the validity of maximum and comparison principles on arbitrary bounded domains. For example, the linear equation $u'' + u = 0$ in one variable will satisfy the maximum principle only on intervals of length less than π. Property (D) plays an essential role in parabolic equations. Condition (N) yields the following elementary but useful fact.

Remark 2.12 (Translations and the negativity condition). For any $\mathcal{F}$ satisfying (N), one has that $u - m$ is $\mathcal{F}$-subharmonic on X for each $m \in (0, +\infty)$ if $u \in \mathrm{USC}(X)$ is $\mathcal{F}$-subharmonic on X. Indeed, for each $x_0 \in \Omega$ and each φ which is C^2 near x_0, one has

$$\varphi \text{ is a test function for } u \text{ at } x_0 \iff \varphi - m \text{ is a test function for } u - m \text{ at } x_0.$$

Conditions (T) and (P) imply the following fact.

Remark 2.13 (Local existence of smooth subharmonics). For any $\mathcal{F}$ satisfying (T), about each point $x_0 \in \mathbb{R}^n$ one can construct strictly $\mathcal{F}$-subharmonic functions which are smooth (and hence bounded) near x_0. Indeed, pick any 2-jet $J_0 := (r_0, p_0, A_0) \in \mathrm{Int}\,\mathcal{F}$, which is nonempty by condition (T). The quadratic polynomial

$$Q(x) := r_0 + \langle p_0, x - x_0 \rangle + \tfrac{1}{2} \langle A_0(x - x_0), x - x_0 \rangle,$$

which has 2-jet $J^2_{x_0} Q = J_0$ at x_0, has 2-jet

$$J^2_x Q = J_0 + (Q(x) - r_0, A_0(x - x_0), 0) \in \mathrm{Int}\,\mathcal{F}$$

for each x sufficiently near to x_0. If, in addition, $\mathcal{F}$ satisfies property (P) then the coherence property of Remark 2.9 implies that Q is $\mathcal{F}$-subharmonic.

Remark 2.14 (Subequations determined by $\mathcal{P}$, $\mathcal{N}$, and $\mathcal{D}$). Each of our three monotonicity conditions (P), (N), and (D) on a subequation $\mathcal{F}$ were phrased as $\mathcal{F} + \mathcal{M} \subset \mathcal{F}$ using the *monotonicity sets* $\mathcal{M}$ defined by

$$\{0\} \times \{0\} \times \mathcal{P}, \quad \mathcal{N} \times \{0\} \times \{0\}, \quad \text{and} \quad \{0\} \times \mathcal{D} \times \{0\},$$

respectively. None of these monotonicity sets are subequations as they violate at least two of the conditions (P), (N), and (T). However, there are three naturally associated subequations for which the sets $\mathcal{P}, \mathcal{N}$, and $\mathcal{D}$ give the *active constraint* while the constraint coming from the other factors is silent, namely

$$\mathcal{M}(\mathcal{P}) := \mathbb{R} \times \mathbb{R}^n \times \mathcal{P}, \quad \mathcal{M}(\mathcal{N}) := \mathcal{N} \times \mathbb{R}^n \times \mathcal{S}(n), \quad \text{and} \quad \mathcal{M}(\mathcal{D}) := \mathbb{R} \times \mathcal{D} \times \mathcal{S}(n).$$
$$(2.15)$$

The second subequation, in isolation, is not very interesting since no derivatives are constrained. However, taken all together these three subequations will be used as the basic building blocks in Chapter 4. It is convenient to refer to $\mathcal{P}$ as the *convexity subequation*, while actually meaning $\mathcal{M}(\mathcal{P}) = \mathbb{R} \times \mathbb{R}^n \times \mathcal{P}$. Similarly, it will be convenient to use $\mathcal{P}$ in place of the monotonicity set $\{0\} \times \{0\} \times \mathcal{P}$ used in the definition of property (P). Similarly, we will refer to $\mathcal{N}$ as the *negativity subequation* and $\mathcal{D}$ as the *$\mathcal{D}$-directional subequation*, while actually meaning $\mathcal{M}(\mathcal{N})$ and $\mathcal{M}(\mathcal{D})$ as in (2.15). This convention that involves reduced constraints will be expanded on and formalized in Convention 3.11 and plays a simplifying role in Chapter 10.

The importance of properties (N) and (T) becomes even more transparent in conjunction with Dirichlet duality for subsets $\mathcal{F} \subset \mathcal{J}^2$, which will be recalled in the next chapter.

Chapter Three

Dirichlet Duality and $\mathcal{F}$-Subharmonic Functions

We begin by recalling the notion of duality introduced in [46] and [49].

Definition 3.1. Suppose $\mathcal{F} \subset \mathcal{J}^2 = \mathbb{R} \times \mathbb{R}^n \times \mathcal{S}(n)$ is an arbitrary subset. The *Dirichlet dual of* $\mathcal{F}$ is defined to be

$$\widetilde{\mathcal{F}} = \sim(-\operatorname{Int}\mathcal{F}) = -(\sim \operatorname{Int}\mathcal{F}), \tag{3.1}$$

where $\sim$ is the complement relative to $\mathcal{J}^2$.

Several elementary properties will be of constant use, and some illustrate the importance of the topological property (T).

Proposition 3.2 (Elementary properties of the Dirichlet dual). *For $\mathcal{F}$, $\mathcal{F}_1$, and $\mathcal{F}_2$ arbitrary subsets of $\mathcal{J}^2$, one has*

(1) $\mathcal{F}_1 \subset \mathcal{F}_2 \;\Rightarrow\; \widetilde{\mathcal{F}}_2 \subset \widetilde{\mathcal{F}}_1$;
(2) $\mathcal{F} + J \subset \mathcal{F} \;\Rightarrow\; \widetilde{\mathcal{F}} + J \subset \widetilde{\mathcal{F}}$ *for each* $J = (r, p, A) \in \mathcal{J}^2$;
(3) $\widetilde{\mathcal{F} + J} = \widetilde{\mathcal{F}} - J$ *for each* $J = (r, p, A) \in \mathcal{J}^2$.

By property (2), one has

(4) $\mathcal{F}$ *satisfies property* (P) $\;\Rightarrow\;$ $\widetilde{\mathcal{F}}$ *satisfies property* (P);
(5) $\mathcal{F}$ *satisfies property* (N) $\;\Rightarrow\;$ $\widetilde{\mathcal{F}}$ *satisfies property* (N);

The topological property (T) *is equivalent to reflexivity, that is,*

(6) $\mathcal{F} = \overline{\operatorname{Int}\mathcal{F}} \;\Leftrightarrow\; \widetilde{\widetilde{\mathcal{F}}} = \mathcal{F}$.

Moreover,

(7) $\mathcal{F}$ *satisfies property* (T) $\;\Rightarrow\;$ $\widetilde{\mathcal{F}}$ *satisfies property* (T);

and hence for $\mathcal{F} \subset \mathcal{J}^2$ a proper subset one has

(8) $\mathcal{F}$ *is a subequation constraint set* $\;\Rightarrow\;$ $\widetilde{\mathcal{F}}$ *is a subequation constraint set.*

Proof. Property (1) follows from definition (3.1) of the dual since $\mathcal{F}_1 \subset \mathcal{F}_2 \Leftrightarrow$ Int $\mathcal{F}_1 \subset$ Int $\mathcal{F}_2$.

For property (2), note that $\mathcal{F} + J \subset \mathcal{F}$ implies that Int $\mathcal{F} + J$ is an open subset of $\mathcal{F}$ and hence Int $\mathcal{F} + J \subset$ Int $\mathcal{F}$, which yields

$$\sim(\text{Int } \mathcal{F}) \subset \sim(\text{Int } \mathcal{F} + J) = (\sim \text{Int } \mathcal{F}) + J \quad \text{and hence } \tilde{\mathcal{F}} \subset \tilde{\mathcal{F}} - J.$$

Thus $\tilde{\mathcal{F}} \supset \tilde{\mathcal{F}} + J$, as desired. The proof of property (3) is similar, using the definitions of $\overline{\mathcal{F} + J}$ and $\tilde{\mathcal{F}}$. Properties (4) and (5) are immediate consequences of (2).

To prove properties (6) and (7), we will use the fact that

$$\text{Int } \mathcal{S} = \sim\overline{(\sim\mathcal{S})} \quad \text{for any subset } \mathcal{S} \subset \mathcal{J}^2. \tag{3.2}$$

For the equivalence (6), using the definition of the dual twice and canceling the minus signs, one has

$$\tilde{\tilde{\mathcal{F}}} = \sim \text{Int}(\sim \text{Int } \mathcal{F}). \tag{3.3}$$

Taking $\mathcal{S} := \sim \text{Int } \mathcal{F}$ in (3.2) transforms (3.3) into

$$\tilde{\tilde{\mathcal{F}}} = \overline{\text{Int } \mathcal{F}},$$

so that $\tilde{\tilde{\mathcal{F}}} = \mathcal{F}$ if and only if property (T) holds for $\mathcal{F}$.

For property (7), first take $\mathcal{S} := \tilde{\mathcal{F}}$ in (3.2) to obtain

$$\text{Int } \tilde{\mathcal{F}} = \sim\overline{(\sim\tilde{\mathcal{F}})} = \sim\overline{(- \text{Int } \mathcal{F})},$$

which, by property (T) for $\mathcal{F}$, proves that Int $\tilde{\mathcal{F}} = \sim(-\mathcal{F})$. Therefore,

$$\overline{\text{Int } \tilde{\mathcal{F}}} = \overline{-(\sim\mathcal{F})} = -\overline{(\sim\mathcal{F})}. \tag{3.4}$$

Next, take $\mathcal{S} := \mathcal{F}$ in (3.2) to obtain $\overline{\sim\mathcal{F}} = \sim(\text{Int } \mathcal{F})$, that is,

$$-\overline{(\sim\mathcal{F})} = -(\sim \text{Int } \mathcal{F}) = \tilde{\mathcal{F}}. \tag{3.5}$$

Combining (3.4) and (3.5) yields $\overline{\text{Int } \tilde{\mathcal{F}}} = \tilde{\mathcal{F}}$.

Finally, property (8) is an immediate consequence of properties (4), (5), and (7) and Definition 2.1. □

It is worth noting that the reverse implications in (4) and (7) may be false. The following is a simple one-dimensional pure second-order example.

Example 3.3. If $\mathcal{F} = \{-1\} \cup [0, +\infty) \subset \mathcal{S}(1)$ then $\widetilde{\mathcal{F}} = [0, +\infty)$ satisfies properties (P) and (T), but $\mathcal{F}$ satisfies neither.

Similarly, the reverse implication in (5) may be false. Next we turn to the behavior of duality under unions and intersections.

Proposition 3.4 (Duality, unions, and intersections). *For $\mathcal{F}_1$ and $\mathcal{F}_2$ arbitrary subsets of $\mathcal{J}^2$, one has*

$$\widetilde{\mathcal{F}_1 \cap \mathcal{F}_2} = \widetilde{\mathcal{F}}_1 \cup \widetilde{\mathcal{F}}_2, \tag{3.6}$$

and if, in addition, $\operatorname{Int}(\mathcal{F}_1 \cup \mathcal{F}_2) = \operatorname{Int}\mathcal{F}_1 \cup \operatorname{Int}\mathcal{F}_2$, *then*

$$\widetilde{\mathcal{F}_1 \cup \mathcal{F}_2} = \widetilde{\mathcal{F}}_1 \cap \widetilde{\mathcal{F}}_2. \tag{3.7}$$

Proof. For the dual of an intersection, note that $\operatorname{Int}(\mathcal{F}_1 \cap \mathcal{F}_2) = (\operatorname{Int}\mathcal{F}_1) \cap (\operatorname{Int}\mathcal{F}_2)$ and use the definition of duality to arrive at (3.6). For the dual of a union, by the definition of duality and the hypothesis on $\operatorname{Int}(\mathcal{F}_1 \cup \mathcal{F}_2)$ one has

$$J \in \widetilde{\mathcal{F}_1 \cup \mathcal{F}_2} \iff -J \notin \operatorname{Int}(\mathcal{F}_1 \cup \mathcal{F}_2) = (\operatorname{Int}\mathcal{F}_1) \cup (\operatorname{Int}\mathcal{F}_2),$$

which yields $J \in \widetilde{F}_1 \cap \widetilde{F}_2$ as desired.   □

It is worth noting that property (3.7) can fail if one does not have the hypothesis $\operatorname{Int}(\mathcal{F}_1 \cup \mathcal{F}_2) = \operatorname{Int}\mathcal{F}_1 \cup \operatorname{Int}\mathcal{F}_2$. The following is a simple pure first-order one-dimensional counterexample.

Example 3.5. In dimension $n = 1$, consider

$$\mathcal{F}_1 := [-1, 0] \quad \text{and} \quad \mathcal{F}_2 := [0, 1].$$

One easily verifies that

$$\widetilde{\mathcal{F}_1 \cup \mathcal{F}_2} = (-\infty, -1] \cup [1, +\infty), \quad \text{while} \quad \widetilde{F}_1 \cap \widetilde{F}_2 = \widetilde{\mathcal{F}_1 \cup \mathcal{F}_2} \cup \{0\}.$$

Additional properties of the dual, the validity of property (T), and algebra of subequations will be addressed in the next chapter on monotonicity (see Propositions 4.5, 4.7, and 4.8).

Notice that property (T) ensures that there is a true duality in the form of the reflexivity property given in (7), and since $\mathcal{F}$ is closed, one also has

$$\partial \mathcal{F} = \mathcal{F} \cap (\sim \operatorname{Int}\mathcal{F}) = \mathcal{F} \cap (-\widetilde{\mathcal{F}}), \tag{3.8}$$

which leads to the following definition.

Definition 3.6. Let $\mathcal{F}$ be a subequation constraint set. A function u is said to be $\mathcal{F}$-*harmonic in* X if

$$u \in \mathcal{F}(X) \quad \text{and} \quad -u \in \widetilde{\mathcal{F}}(X). \tag{3.9}$$

A function u is said to be $\mathcal{F}$-*superharmonic in* X if $-u \in \widetilde{\mathcal{F}}(X)$.

Remark 3.7. An equivalent formulation of $u \in \mathrm{LSC}(\Omega)$ being $\mathcal{F}$-superharmonic in X is that for each $x_0 \in \Omega$, one has

$$J^{2,-}_{x_0} u \subset {\sim} \operatorname{Int} \mathcal{F}, \tag{3.10}$$

where $J^{2,-}_{x_0} u$ is the set of (*lower*) *test jets for* u *at* x_0, that is, the set of 2-jets $J^2_{x_0}\varphi$ with φ a C^2 (*lower*) *test function for* u *at* x_0 (φ is C^2 near x_0 and $u - \varphi$ has a local minimum value of zero in x_0).

Notice that $\mathcal{F}$-harmonic functions are automatically continuous and that $\mathcal{F}$-superharmonic functions are automatically lower semicontinuous. Moreover, $u \in C^2(X)$ is $\mathcal{F}$-harmonic on X if and only if

$$J_x u \in \partial \mathcal{F} \quad \text{for each } x \in X, \tag{3.11}$$

which follows from the coherence property for $\mathcal{F}$, $\widetilde{\mathcal{F}}$ of Remark 2.9 coupled with (3.9) and (3.8).

Example 3.8 (Classical subharmonic and harmonic functions). If $\mathcal{F} := \mathbb{R} \times \mathbb{R}^n \times \mathcal{F}_\Delta$ with $\mathcal{F}_\Delta = \{A \in \mathcal{S}(n) : \operatorname{tr} A \geq 0\}$, then $\widetilde{\mathcal{F}} = \mathcal{F}$ is self-dual and $\mathcal{F}$-harmonics are characterized by satisfying the mean value property, that is, by having equality in (2.11).

Pairs of $\mathcal{F}$ and $\widetilde{\mathcal{F}}$-subharmonics are the key players in the comparison principle when making use of Dirichlet duality, and hence a few additional pairs are worth mentioning now.

Example 3.9 (Convex and subaffine functions). If $\mathcal{F} := \mathbb{R} \times \mathbb{R}^n \times \mathcal{P}$ is the *convexity subequation*, then $\widetilde{\mathcal{F}} = \mathbb{R} \times \mathbb{R}^n \times \widetilde{\mathcal{P}}$, where $\widetilde{\mathcal{P}} = {\sim}(-\operatorname{Int}\mathcal{P})$ is the complement of the negative definite quadratic forms. Thus $A \in \widetilde{\mathcal{P}} \Leftrightarrow \lambda_{\max}(A) \geq 0$, where $\lambda_{\max}$ is the maximal eigenvalue. The class of $\widetilde{\mathcal{F}}$-subharmonic functions is characterized as follows: $u \in \mathrm{USC}(X)$ belongs to $\widetilde{\mathcal{F}}(X)$ if and only if for every open subset Ω with $\Omega \subset\subset X$ and each affine function a, one has

$$u \leq a \text{ on } \partial\Omega \;\Rightarrow\; u \leq a \text{ on } \Omega. \tag{3.12}$$

Since they are equal (see [46, Proposition 4.5]), we will denote this space of *subaffine functions on* X by $\mathrm{SA}(X)$ as well as $\widetilde{\mathcal{F}}(X)$.

Example 3.10 (Negative functions). If $\mathcal{F} := \mathcal{N} \times \mathbb{R}^n \times \mathcal{S}(n)$, then $\widetilde{\mathcal{F}} = \mathcal{F}$ is self-dual and zeroth order.

Convention 3.11. If the constraint $\mathcal{F}$ places restrictions only on the matrix variable, that is, if $\mathcal{F} := \mathbb{R} \times \mathbb{R}^n \times \mathcal{F}_2$ with $\mathcal{F}_2$ satisfying (P): $\mathcal{F}_2 + \mathcal{P} \subset \mathcal{F}_2$, then "$u$ is $\mathcal{F}$-subharmonic on X" and "u is $\mathcal{F}_2$-subharmonic on X" will mean the same thing. For example, we can denote by $\mathcal{P}(X)$ the convex functions of (2.10) and by $\widetilde{\mathcal{P}}(X)$ the subaffine functions of Example 3.9. Similarly, the negative functions of Example 3.10 can be denoted by $\mathcal{N}(X)$. In addition, when it is clear from the context, we will refer to the *subequation* $\mathcal{F}_2$ meaning $\mathcal{F}$.

Additional justification for this convention is provided in the discussion before Remark 10.1.

Example 3.12 (Negative convex and subaffine-plus functions). If $\mathcal{F} := \mathcal{N} \times \mathbb{R}^n \times \mathcal{P}$ is the *negativity and convexity subequation*, then

$$\widetilde{\mathcal{F}} = (\mathcal{N} \times \mathbb{R}^n \times \mathcal{S}(n)) \cup (\mathbb{R} \times \mathbb{R}^n \times \widetilde{\mathcal{P}}) \tag{3.13}$$

and the dual subharmonics are characterized by

$$w \in \widetilde{\mathcal{F}}(X) \quad \Leftrightarrow \quad w^+ := \max\{w, 0\} \in \widetilde{\mathcal{P}}(X) = \mathrm{SA}(X) \tag{3.14}$$

and referred to as the *subaffine-plus functions on X*. This characterization is discussed in Theorem 10.7. Denoting $\mathcal{Q} = \mathcal{N} \times \mathcal{P}$ and $\widetilde{\mathcal{Q}} := \{(r, A) \in \mathbb{R} \times \mathcal{S}(n) : r \in \mathcal{N}$ or $A \in \widetilde{\mathcal{P}}\}$, we will also say that the subaffine-plus functions are $\widetilde{\mathcal{Q}}$-subharmonic on X in the spirit of Convention 3.11.

Additional examples include $\mathcal{F} = \mathcal{M}$ where $\mathcal{M}$ is a *monotonicity cone*, as will be discussed in the next chapter. For future reference we record the following example of duals to the elementary monotonicity cones introduced in (2.15).

Example 3.13. The duals of

$$\mathcal{M}(\mathcal{P}) := \mathbb{R} \times \mathbb{R}^n \times \mathcal{P}, \quad \mathcal{M}(\mathcal{N}) := \mathcal{N} \times \mathbb{R}^n \times \mathcal{S}(n), \quad \text{and} \quad \mathcal{M}(\mathcal{D}) := \mathbb{R} \times \mathcal{D} \times \mathcal{S}(n)$$

are

$$\widetilde{\mathcal{M}}(\mathcal{P}) = \mathbb{R} \times \mathbb{R}^n \times \widetilde{\mathcal{P}}, \quad \widetilde{\mathcal{M}}(\mathcal{N}) = \mathcal{N} \times \mathbb{R}^n \times \mathcal{S}(n), \quad \text{and} \quad \widetilde{\mathcal{M}}(\mathcal{D}) = \mathbb{R} \times \widetilde{\mathcal{D}} \times \mathcal{S}(n), \tag{3.15}$$

as follows easily from applying definition (3.1) of duality.

We conclude this chapter with a very useful result which illustrates the importance of the negativity condition (N), saying that the comparison principle always holds if the function v is C^2-*smooth* and *strictly* $\widetilde{\mathcal{F}}$-subharmonic.

Lemma 3.14 (Definitional comparison). *Suppose that $\mathcal{F}$ is a subequation constraint set and that $u \in \mathrm{USC}(X)$.*

(a) *If u is $\mathcal{F}$-subharmonic on X, then the following form of the comparison principle holds for each bounded domain $\Omega \subset\subset X$:*

$$\begin{cases} u + v \le 0 \text{ on } \partial\Omega \;\Rightarrow\; u + v \le 0 \text{ on } \Omega \\ \text{if } v \in \mathrm{USC}(\overline{\Omega}) \cap C^2(\Omega) \text{ is strictly } \widetilde{\mathcal{F}}\text{-subharmonic on } X. \end{cases} \tag{3.16}$$

With $w := -v$ one has the equivalent statement

$$\begin{cases} u \le w \text{ on } \partial\Omega \;\Rightarrow\; u \le w \text{ on } \Omega \\ \text{if } w \in \mathrm{LSC}(\overline{\Omega}) \cap C^2(\Omega) \text{ with } J^2_x w \notin \mathcal{F} \text{ for each } x \in \Omega \end{cases} \tag{3.17}$$

(that is, for w which are regular and strictly $\mathcal{F}$-superharmonic in X).
(b) *Conversely, suppose that for each $x \in X$ there are arbitrarily small neighborhoods $\Omega \subset\subset X$ about x where the form of comparison of part (a) holds. Then u is $\mathcal{F}$-subharmonic on X.*

Proof. Suppose that (3.17) fails for some domain $\Omega \subset\subset X$ and some regular strictly $\mathcal{F}$-superharmonic function w. Then $u - w \in \mathrm{USC}(\overline{\Omega})$ will have a positive maximum value $m > 0$ at an interior point $x_0 \in \Omega$ and hence $\varphi := w + m$ is C^2 near x_0 and satisfies

$$u - \varphi \le 0 \quad \text{near } x_0, \text{ with equality at } x_0.$$

Since u is $\mathcal{F}$-subharmonic at x_0, by Definition 2.4 one has

$$J^2_{x_0}(w + m) = J^2_{x_0} w + (m, 0, 0) \in \mathcal{F},$$

which contradicts property (N) since $m > 0$ and $J^2_{x_0} w \notin \mathcal{F}$. This completes the proof of part (a).

For part (b), suppose that u fails to be $\mathcal{F}$-subharmonic at some $x_0 \in X$. By the bad test jet lemma (Lemma 2.8), there exist $\rho, \varepsilon > 0$ and $J = (u(x_0), p, A) \notin \mathcal{F}$ such that

$$u(x) - Q_J(x) \le -\varepsilon|x - x_0|^2 \quad \text{on } \overline{B}_\rho(x_0), \text{ with equality at } x_0, \tag{3.18}$$

where Q_J is the quadratic with $J^2_{x_0} Q_J = J$. Notice that (3.18) continues to hold for all smaller choices of $\rho > 0$ and $\varepsilon > 0$. The function $-Q_J$ is smooth and strictly $\widetilde{\mathcal{F}}$-subharmonic in x_0 since $J^2_{x_0}(-Q_J) = -J \in -(\sim\mathcal{F}) = \mathrm{Int}\,\widetilde{\mathcal{F}}$. Since $\mathrm{Int}\,\widetilde{\mathcal{F}}$ is open, by choosing $\rho > 0$ and $\varepsilon > 0$ sufficiently small, the function $v := -Q_J + \varepsilon\rho^2$ will be strictly $\widetilde{\mathcal{F}}$-subharmonic in $B_\rho(x_0)$. Since

$$u + v = 0 \text{ on } \partial B_\rho(x_0) \quad \text{and} \quad u(x_0) + v(x_0) = \varepsilon\rho^2 > 0,$$

the comparison (3.16) fails for v on $\overline{B}_\rho(x_0)$, with $\rho > 0$ arbitrarily small. $\quad\square$

Remark 3.15. Note that the above proof is very general. The result holds verbatim on any manifold X and any subequation $\mathcal{F}$ on X. More precisely, it holds if $\mathcal{F}$ satisfies properties (P) and (N), and $\mathcal{F} = \overline{\text{Int}\,\mathcal{F}}$. Replacing $\mathcal{F}$ by its dual $\widetilde{\mathcal{F}}$, since $\widetilde{\widetilde{\mathcal{F}}} = \mathcal{F}$ (the proof of reflexivity makes use of the definition of duality and (3.2), which holds in any topological space), one also has the comparison principle (3.16) if the function u is C^2-*smooth* and *strictly* $\mathcal{F}$-subharmonic and $v \in \text{USC}(X)$ is $\widetilde{\mathcal{F}}$-subharmonic.

We note that Lemma 3.14 will be useful in the proof (in Chapter 6) of the *zero maximum principle* for $\widetilde{\mathcal{M}}(X)$-subharmonic functions, where $\mathcal{M}$ is a *monotonicity cone*, which is the subject of Chapter 4.

Remark 3.16 (A viscosity "tool kit" for $\mathcal{F}$-subharmonic functions). The definitional comparison lemma (Lemma 3.14) represents one aspect of the "nuts and bolts" of handling $\mathcal{F}$-subharmonic functions (viscosity subsolutions). Other such tools have been discussed previously, such as the bad test jet lemma (Lemma 2.8) and the almost everywhere theorem (Lemma 2.10). In addition, one has the subharmonic addition theorem (Theorem 7.1) and the elementary properties of $\mathcal{F}$-subharmonic functions of Proposition D.1.

We conclude this chapter by adding a useful classical computational lemma to our tool kit, for the reader's convenience and future reference. We combine the statement and the proofs as a remark. First, we need some notation. For $x \neq 0$ in $\mathbb{R}^n$, we denote orthogonal projection onto the one-dimensional linear subspace $[x]$ through x by

$$P_x := \frac{1}{|x|^2} x \otimes x, \tag{3.19}$$

and similarly, by

$$P_{x^\perp} := I - \frac{1}{|x|^2} x \otimes x = I - P_x, \tag{3.20}$$

the projection onto the orthogonal complement $[x]^\perp$.

Remark 3.17 (Radial calculations and examples). First, note that for C^2-functions $v \colon \Omega \to \mathbb{R}$ and $\psi \colon v(\Omega) \subset \mathbb{R} \to \mathbb{R}$, the chain rule applied to $u = \psi \circ v \in C^2(\Omega)$ gives (for each $x \in \Omega$)

$$Du(x) = D_x(\psi(v(x))) = \psi'(v(x))Dv(x) \tag{3.21}$$

and

$$D^2u(x) = D_x^2(\psi(v(x))) = \psi'(v(x))D^2v(x) + \psi''(v(x))Dv(x) \otimes Dv(x). \tag{3.22}$$

In particular, with $v(x) = |x|$ and ψ of class C^2 on an interval in $(0, +\infty)$, since

$$Dv(x) = D_x(|x|) = \frac{x}{|x|} \quad \text{and} \quad D^2 v(x) = D_x^2(|x|) = \frac{1}{|x|} P_{x^\perp}, \qquad (3.23)$$

the radial function defined by $u(x) := \psi(|x|)$ has reduced 2-jet $(Du(x), D^2u(x))$, given by

$$D_x \psi(|x|) = \psi'(|x|) \frac{x}{|x|} \quad \text{and} \quad D_x^2 \psi(|x|) = \frac{\psi'(|x|)}{|x|} P_{x^\perp} + \psi''(|x|) P_x, \qquad (3.24)$$

on the corresponding annular domain. In addition to the example $\psi(t) := t$ in (3.23), here are some other useful examples.

(1) For $u(x) := \frac{1}{2}|x|^2$, that is, $\psi(t) := \frac{1}{2}t^2$, one has

$$(Du(x), D^2u(x)) = (x, I). \qquad (3.25)$$

Examples (3.23) and (3.25) generalize from $m = 0$ and $m = 1$ to the following:

(2) For $u(x) := \frac{1}{m+1}|x|^{m+1}$, that is, $\psi(t) := \frac{1}{m+1}t^{m+1}$, with $m \in \mathbb{R}$, $m \neq 1$, one has

$$\begin{aligned}(Du(x), D^2u(x)) &= |x|^{m-1}(x, P_{x^\perp} + mP_x) \\ &= |x|^{m-1}(x, I + (m-1)P_x).\end{aligned} \qquad (3.26)$$

(3) For $u(x) := e^{\alpha|x|^2/2}$, that is, $\psi(t) := e^{\alpha t^2/2}$, with $\alpha \in \mathbb{R}$, one has

$$\begin{aligned}(Du(x), D^2u(x)) &= \alpha e^{\alpha \frac{|x|^2}{2}}(x, P_{x^\perp} + (\alpha+1)P_x) \\ &= \alpha e^{\alpha \frac{|x|^2}{2}}(x, I + \alpha P_x).\end{aligned} \qquad (3.27)$$

These two examples $\frac{1}{m+1}|x|^{m+1}$ and $e^{\alpha|x|^2/2}$ can usually be used interchangeably, since up to a positive scalar multiple, the reduced 2-jet is $(x, I + (m-1)P_x)$ or $(x, I + \alpha P_x)$.

(4) For $u(x) := e^{\alpha|x|}$, that is, $\psi(t) := e^{\alpha t}$ with $\alpha \in \mathbb{R}$, one has

$$\begin{aligned}(Du(x), D^2u(x)) &= \frac{\alpha}{|x|} e^{\alpha|x|}(x, P_{x^\perp} + \alpha|x| P_x) \\ &= \frac{\alpha}{|x|} e^{\alpha|x|}(x, I + (\alpha|x| - 1)P_x).\end{aligned} \qquad (3.28)$$

Chapter Four

Monotonicity Cones for

Constant-Coefficient Subequations

Definition 4.1. Given a subset $\mathcal{F} \subset \mathcal{J}^2$, a set $\mathcal{M} \subset \mathcal{J}^2$ is called a *monotonicity set for* $\mathcal{F}$ and we say $\mathcal{F}$ *is* $\mathcal{M}$-*monotone* if

$$\mathcal{F} + \mathcal{M} \subset \mathcal{F}. \tag{4.1}$$

Since

$$\mathcal{F} \text{ satisfies (P) and (N)} \quad \Leftrightarrow \quad \mathcal{F} \text{ is } \mathcal{N} \times \{0\} \times \mathcal{P}\text{-monotone,} \tag{4.2}$$

the set

$$\mathcal{M}_0 := \mathcal{N} \times \{0\} \times \mathcal{P} \tag{4.3}$$

will be called the *minimal monotonicity set for all subequation constraint sets* $\mathcal{F}$.

This set is a closed convex cone which satisfies properties (P) and (N), but it is *not* a subequation since property (T) fails, or equivalently Int $\mathcal{M}_0 = \emptyset$.

A larger monotonicity set $\mathcal{M}$ for $\mathcal{F}$ provides more information about $\mathcal{F}$. Note that the sum $\mathcal{M}_1 + \mathcal{M}_2$ of two monotonicity sets for $\mathcal{F}$ is again a monotonicity set for $\mathcal{F}$. Hence, if $\mathcal{F}$ is a subequation, we can always add $\mathcal{M}_0$ to $\mathcal{M}$, since $\mathcal{F} + \mathcal{M}_0 \subset \mathcal{F}$. Also, we can replace $\mathcal{M}$ by its closure (assuming that $\mathcal{F}$ is closed) and the resulting set will still be a monotonicity set for $\mathcal{F}$. Said differently, we need only consider monotonicity sets $\mathcal{M}$ for subequation $\mathcal{F}$ with the properties

$$\text{(i) } \mathcal{M}_0 \subset \mathcal{M}, \quad \text{(ii) } \mathcal{M} + \mathcal{M} \subset \mathcal{M}, \quad \text{and (iii) } \mathcal{M} \text{ is closed.} \tag{4.4}$$

For simplicity, in this book we restrict attention to sets $\mathcal{M}$ which are *cones*, that is,

$$t\mathcal{M} \subset \mathcal{M} \quad \text{for each } t \geq 0. \tag{4.5}$$

All cones are taken to have vertex at the origin, unless stated otherwise. Note that

$$\text{if } \mathcal{M} \text{ is a cone then } \mathcal{M} + \mathcal{M} \text{ is a convex cone.} \tag{4.6}$$

As a consequence of (4.4) and (4.5), if $\mathcal{M}$ is a monotonicity cone for a subequa-
tion $\mathcal{F}$ then $\overline{C(\mathcal{M})}$, the closed convex cone hull of $\mathcal{M}$, is also a monotonicity
cone for $\mathcal{F}$. Said differently, given a subequation $\mathcal{F}$, a monotonicity cone $\mathcal{M}$ for
$\mathcal{F}$ can always be enlarged to one where

$$\text{(i) } \mathcal{M}_0 \subset \mathcal{M} \quad \text{and} \quad \text{(ii) } \mathcal{M} \text{ is closed convex cone.} \tag{4.7}$$

4.1 THE MAXIMAL MONOTONICITY CONE SUBEQUATION

Among all sets $\mathcal{M}$ which are cones and for which a given subequation $\mathcal{F}$ is
$\mathcal{M}$-monotone, there is a unique largest or maximal monotonicity cone for $\mathcal{F}$,
defined as follows.

Definition 4.2. Suppose that $\mathcal{F} \subset \mathcal{J}^2$ is a subequation. Associated to $\mathcal{F}$ is its
maximal monotonicity cone $\mathcal{M}_\mathcal{F}$ defined by

$$\mathcal{M}_\mathcal{F} := \left\{ J \in \mathcal{J}^2 : \mathcal{F} + tJ \subset \mathcal{F} \text{ for each } t \in [0,1] \right\}. \tag{4.8}$$

This invariant satisfies (4.7) and is characterized in the following result.

Proposition 4.3. *Suppose that $\mathcal{F} \subset \mathcal{J}^2$ is a subequation. Its maximal monotonicity cone $\mathcal{M}_\mathcal{F}$ is a closed convex cone containing the minimal monotonicity set
$\mathcal{M}_0 = \mathcal{N} \times \{0\} \times \mathcal{P}$ and hence satisfies properties (N) and (P). Therefore,*

$$\mathcal{M}_\mathcal{F} \text{ is a subequation } \Leftrightarrow \mathcal{M}_\mathcal{F} \text{ satisfies the topological property (T).} \tag{4.9}$$

*Moreover, since $\mathcal{M}_\mathcal{F}$ is a closed convex cone, (T) is satisfied if and only if
$\text{Int} \, \mathcal{M}_\mathcal{F} \neq \emptyset$. Finally, $\mathcal{M}_\mathcal{F}$ is maximal in the sense that if $\mathcal{M}$ is a cone and is a
monotonicity set for $\mathcal{F}$, then*

$$\mathcal{M} \subset \mathcal{M}_\mathcal{F}. \tag{4.10}$$

Proof. It is easy to see that $\mathcal{F}$ is closed implies that $\mathcal{M}_\mathcal{F}$ is closed. By (4.2), $\mathcal{F}$
satisfies (P) and (N) if and only if $\mathcal{N} \times \{0\} \times \mathcal{P} \subset \mathcal{M}_\mathcal{F}$. Obviously, $\mathcal{M}_\mathcal{F} + \mathcal{M}_\mathcal{F} \subset$
$\mathcal{M}_\mathcal{F}$ and hence if $J \in \mathcal{M}_\mathcal{F}$ then each integer multiple $kJ \in \mathcal{M}_\mathcal{F}$. By the definition
of $\mathcal{M}_\mathcal{F}$ this implies that $tJ \in \mathcal{M}_\mathcal{F}$ for each $t \in [0,k]$, which proves that $\mathcal{M}_\mathcal{F}$ is a
cone. Since $\mathcal{M}_\mathcal{F} + \mathcal{M}_\mathcal{F} \subset \mathcal{M}_\mathcal{F}$, this implies that $\mathcal{M}_\mathcal{F}$ is a convex cone.

The statements regarding property (T) are straightforward. The final maxi-
mality claim for $\mathcal{M}_\mathcal{F}$ is immediate from the definition of $\mathcal{M}_\mathcal{F}$. $\square$

Remark 4.4. In light of Proposition 4.3, given a subequation $\mathcal{F} \subset \mathcal{J}^2$, the question of whether or not $\mathcal{F}$ has a monotonicity cone which is a subequation reduces

to (is equivalent to) the following question:

Does the maximal monotonicity cone $\mathcal{M}_{\mathcal{F}}$ for $\mathcal{F}$ have interior?   (4.11)

As we will see in what follows, in order to apply the techniques of this book to a subequation $\mathcal{F}$, it must have monotonicity $\mathcal{M}$ which is a subequation.

Some of the additional properties of the maximal monotonicity cone $\mathcal{M}_{\mathcal{F}}$ of a subequation $\mathcal{F} \subset \mathcal{J}^2$ are as follows.

Proposition 4.5.

(a) *If a subequation $\mathcal{F}$ is a convex cone, then $\mathcal{M}_{\mathcal{F}} = \mathcal{F}$. In particular, if $\mathcal{M}_{\mathcal{F}}$ is also a subequation, then $\mathcal{M}_{\mathcal{M}_{\mathcal{F}}} = \mathcal{M}_{\mathcal{F}}$.*
(b) *A subequation $\mathcal{F}$ and its dual $\widetilde{\mathcal{F}}$ have the same maximal monotonicity cones, that is, $\mathcal{M}_{\widetilde{\mathcal{F}}} = \mathcal{M}_{\mathcal{F}}$.*

Proof. Part (a) follows from the fact that if $\mathcal{C}$ is any convex cone in $\mathbb{R}^N$, then for each $v \in \mathbb{R}^N$,
$$\mathcal{C} + v \subset \mathcal{C} \iff v \in \mathcal{C}.$$

Part (b) follows from the fact for any subequation $\mathcal{F}$, one has
$$\text{for each } J \in \mathcal{J}^2 : \quad \mathcal{F} + J \subset \mathcal{F} \iff \widetilde{\mathcal{F}} + J \subset \widetilde{\mathcal{F}}. \tag{4.12}$$

Indeed, the implication ($\Rightarrow$) is part (2) of Proposition 3.2. By the same property, if $\widetilde{\mathcal{F}} + J \subset \widetilde{\mathcal{F}}$, then $\widetilde{\widetilde{\mathcal{F}}} + J \subset \widetilde{\widetilde{\mathcal{F}}}$, but $\widetilde{\widetilde{\mathcal{F}}} = \mathcal{F}$ by the reflexivity in part (6) of Proposition 3.2, since any subequation $\mathcal{F}$ satisfies property (T).   □

We record the following basic definitions.

Definition 4.6. A closed convex cone $\mathcal{M} \subset \mathcal{J}^2$ with vertex at the origin that contains the minimal monotonicity set $\mathcal{M}_0 := \mathcal{N} \times \{0\} \times \mathcal{P}$ will be referred to as a *monotonicity cone*. A monotonicity cone $\mathcal{M}$ which satisfies property (T) will be called a *monotonicity cone subequation*.

In this case, for any closed set $\mathcal{F} \subset \mathcal{J}$ which is $\mathcal{M}$-monotone, property (T) for $\mathcal{F}$ follows.

Proposition 4.7. *Let $\mathcal{M}$ be a monotonicity cone subequation. Then, for any closed subset $\mathcal{F} \subset \mathcal{J}^2$ which is $\mathcal{M}$-monotone, one has the set identities*
$$\mathcal{F} + \operatorname{Int} \mathcal{M} = \operatorname{Int} \mathcal{F} \tag{4.13}$$

and
$$\mathcal{F} = \overline{\operatorname{Int} \mathcal{F}}, \tag{4.14}$$

and hence $\mathcal{F}$ satisfies properties (P), (N), *and* (T), *so that $\mathcal{F}$ is a subequation constraint set (if $\mathcal{F} \neq \emptyset$ and $\mathcal{F} \neq \mathcal{J}^2$).*

Proof. For identity (4.13), first note that $\mathcal{F} + \operatorname{Int} \mathcal{M}$ is an open set, being the union over $J \in \mathcal{F}$ of the open sets $J + \operatorname{Int} \mathcal{M}$. By monotonicity, this open set is contained in $\mathcal{F}$ and hence is contained in $\operatorname{Int} \mathcal{F}$. For the reverse inclusion, if $J \in \operatorname{Int} \mathcal{F}$ then, picking $J_0 \in \operatorname{Int} \mathcal{M}$, one has

$$J = (J - \varepsilon J_0) + \varepsilon J_0 \in \operatorname{Int} \mathcal{F} + \operatorname{Int} \mathcal{M},$$

if $\varepsilon > 0$ is chosen sufficiently small, since $\operatorname{Int} \mathcal{F}$ is open and $\operatorname{Int} \mathcal{M}$ is a cone.

For identity (4.14), with $J_0 \in \operatorname{Int} \mathcal{M}$, each $J \in \mathcal{F}$ can be approximated by $J + \varepsilon J_0$, which belongs to $\operatorname{Int} \mathcal{F}$ by identity (4.13).

The subequation claim for $\mathcal{F}$ is immediate, since (4.14) is the topological property (T) and the properties (P) and (N) follow from the $\mathcal{M}_0$-monotonicity of $\mathcal{F}$. □

As a corollary of the fact that monotonicity implies the topological property (T), we have the following result.

Proposition 4.8 (Intersections and unions). *Suppose that $\{\mathcal{F}_\sigma : \sigma \in \Sigma\}$ is an arbitrary family of subequations which are all $\mathcal{M}$-monotone for a given monotonicity subequation cone $\mathcal{M}$. Then one has the following statements:*

(a) *The intersection $\mathcal{F} := \bigcap_{\sigma \in \Sigma} \mathcal{F}_\sigma$ (if nonempty) is an $\mathcal{M}$-monotone subequation.*
(b) *The closure of the union $\mathcal{F} := \overline{\bigcup_{\sigma \in \Sigma} \mathcal{F}_\sigma}$ (if not equal to all of $\mathcal{J}^2$) is an $\mathcal{M}$-monotone subequation.*

Proof. First, note that arbitrary intersections and arbitrary unions of sets that are $\mathcal{M}$-monotone are again $\mathcal{M}$-monotone for any set $\mathcal{M}$. In particular, property (P), as well as property (N), is preserved under arbitrary intersections and unions. In addition, if $\mathcal{M}$ is a monotonicity cone subequation then $\mathcal{M}$-monotonicity for a set $\mathcal{F}$ implies property (T) for $\mathcal{F}$ as long as $\mathcal{F}$ is closed (see Proposition 4.7). Arbitrary intersections of closed sets are closed, while only finite unions of closed sets are closed in general (which is why in general one must take the closure of the union). Finally, subequations $\mathcal{F} \subset \mathcal{J}^2$ are by definition nonempty and proper subsets of $\mathcal{J}^2$. Intersections of proper subsets are proper, but may be empty. Unions of nonempty sets are nonempty but may have closure equal to all of $\mathcal{J}^2$. This completes the proof. □

It is worth noting that easy examples (including pure first-order subequations where (P) and (N) are automatic) illustrate the role monotonicity plays in Proposition 4.8. More precisely, for arbitrary closed sets, property (T) is not preserved under finite intersections and the interior of finite unions may be larger than the (finite) union of the interiors. For example, consider two closed cubes of the same size which intersect along a common face.

Now we turn to examples of monotonicity cone subequations.

4.2   PRODUCT MONOTONICITY CONE SUBEQUATIONS

Consider a product set

$$\mathcal{M} := \mathcal{M}^0 \times \mathcal{M}^1 \times \mathcal{M}^2 \subset \mathcal{J}^2 = \mathbb{R} \times \mathbb{R}^n \times \mathcal{S}(n).$$

Note that $\mathcal{M}$ is a convex cone if and only if each factor of $\mathcal{M}$ is a convex cone, and $\mathcal{M}$ satisfies (T) if and only if each factor does; that is,

$$\mathcal{M} = \overline{\mathrm{Int}\,\mathcal{M}} \iff \mathcal{M}^k = \overline{\mathrm{Int}\,\mathcal{M}^k}, \quad k = 0, 1, 2.$$

In this case, $\mathcal{M}^1$ is a convex cone satisfying property (T). Thus,

$$\mathcal{M}^1 := \mathcal{D} \text{ is a directional cone as in Definition 2.2,}$$

that is, a closed convex cone in $\mathbb{R}^n$ with vertex at the origin and nonempty interior. Note that $\mathcal{D} = \mathbb{R}^n$ is allowed. Also note that

$$\mathcal{M} \text{ satisfies (P)} \iff \mathcal{P} \subset \mathcal{M}^2 \tag{4.15}$$

and that

$$\mathcal{M} \text{ satisfies (N)} \iff \mathcal{N} \subset \mathcal{M}^0. \tag{4.16}$$

This can be summarized as follows.

Proposition 4.9. *A product set* $\mathcal{M} = \mathcal{M}^0 \times \mathcal{M}^1 \times \mathcal{M}^2 \subset \mathcal{J}^2$ *is a monotonicity cone subequation if and only if the factors satisfy*

(0) $\mathcal{M}^0 = \mathcal{N}$ *or* $\mathcal{M}^0 = \mathbb{R}$;
(1) $\mathcal{M}^1 = \mathcal{D}$ *is a directional cone in* $\mathbb{R}^n$ *as in Definition 2.2;*
(2) $\mathcal{M}^2$ *is a closed convex cone in* $\mathcal{S}(n)$ *which contains* $\mathcal{P}$.

In particular, important examples of product monotonicity cone subequations include those determined by $\mathcal{N}$, $\mathcal{D}$, and $\mathcal{P}$ as introduced in (2.15), that is,

$$\mathcal{M}(\mathcal{P}) := \mathbb{R} \times \mathbb{R}^n \times \mathcal{P}, \quad \mathcal{M}(\mathcal{N}) := \mathcal{N} \times \mathbb{R}^n \times \mathcal{S}(n), \quad \text{and} \quad \mathcal{M}(\mathcal{D}) := \mathbb{R} \times \mathcal{D} \times \mathcal{S}(n).$$

Moreover, each product monotonicity cone subequation contains the intersection of these basic examples; that is, these intersections form a fundamental neighborhood system for $\mathcal{M}_0$ among all product monotonicity subequations. The proof is omitted.

Definition 4.10. A *fundamental product monotonicity subequation* is a subset $\mathcal{M} \subset \mathcal{J}^2$ of the form

$$\mathcal{M} := \mathcal{N} \times \mathcal{D} \times \mathcal{P} = \mathcal{M}(N) \cap \mathcal{M}(\mathcal{D}) \cap \mathcal{M}(\mathcal{P}), \qquad (4.17)$$

determined by the choice of a directional cone $\mathcal{D}$ in $\mathbb{R}^n$.

An arbitrary subset $\mathcal{F} \subset \mathcal{J}^2$ is $\mathcal{M}$-monotone for a fundamental product monotonicity subequation $\mathcal{M}$ as in (4.17) if and only if $\mathcal{F}$ satisfies (P), (N), and the *directionality property*

$$(r, p, A) \in \mathcal{F} \;\Rightarrow\; (r, p+q, A) \in \mathcal{F} \quad \text{for all } q \in \mathcal{D}. \qquad (\text{D})$$

Lemma 4.11. *The Dirichlet dual of a fundamental product monotonicity subequation $\mathcal{M}$ is*

$$\widetilde{\mathcal{M}} = \widetilde{\mathcal{M}}(N) \cup \widetilde{\mathcal{M}}(\mathcal{D}) \cup \widetilde{\mathcal{M}}(\mathcal{P})$$
$$= \{(r, p, A) \in \mathcal{J}^2 : \textit{either } r \leq 0, \textit{ or } p \in \widetilde{\mathcal{D}}, \textit{ or } A \in \widetilde{\mathcal{P}}\}. \qquad (4.18)$$

Proof. The first line of (4.18) just uses the formula for the dual of intersections (3.6) and the second line uses the formulas (3.15) for the duals of the elementary cones. $\qquad\square$

Chapter Five

A Fundamental Family of Monotonicity Cone Subequations

In this chapter we will construct a family of monotonicity cone subequations which is fundamental in the sense that if a given subequation $\mathcal{F}$ is $\mathcal{M}$-monotone for a monotonicity subequation $\mathcal{M}$, then there exists an element $\mathcal{M}_*$ of the fundamental family with $\mathcal{M}_* \subset \mathcal{M}$. That is, the family provides a fundamental neighborhood system for $\mathcal{M}_0$ among all convex cone subequations. Note that $\mathcal{M}_* \subset \mathcal{M}$ implies $\widetilde{\mathcal{M}} \subset \widetilde{\mathcal{M}}_*$, so that if the (ZMP) holds for $\widetilde{\mathcal{M}}_*$ it will also hold for $\widetilde{\mathcal{M}}$.

In order to construct the fundamental family, we will use intersections of 5 elementary monotonicity cone subequations to build up a family of 17 distinct monotonicity cone subequations for use in the main comparison theorem of Chapter 7. Some of these will also have a product structure and in some cases there will be silent factors which are suitable for the results on reduction of Chapter 10.

5.1 CONSTRUCTION OF THE FUNDAMENTAL FAMILY

We begin by adding two one-parameter families of monotonicity cones to the elementary building blocks $\mathcal{M}(\mathcal{P})$, $\mathcal{M}(\mathcal{N})$, and $\mathcal{M}(\mathcal{D})$ given in (2.15) as

$$\mathcal{M}(\mathcal{P}) := \mathbb{R} \times \mathbb{R}^n \times \mathcal{P}, \quad \mathcal{M}(\mathcal{N}) := \mathcal{N} \times \mathbb{R}^n \times \mathcal{S}(n), \quad \text{and} \quad \mathcal{M}(\mathcal{D}) := \mathbb{R} \times \mathcal{D} \times \mathcal{S}(n),$$

where

$$\mathcal{P} = \{A \in \mathcal{S}(n) : A \geq 0\}, \quad \mathcal{N} = (-\infty, 0] \subset \mathbb{R}, \quad \text{and} \quad \mathcal{D} \subset \mathbb{R}^N,$$

and $\mathcal{D}$ is a closed convex cone with vertex in 0 and $\text{Int } \mathcal{D} \neq \emptyset$.

Definition 5.1. Let $\mathcal{J} = \mathbb{R} \times \mathbb{R}^n \times \mathcal{S}(n)$. Given a real number $\gamma \in (0, +\infty)$, the set

$$\mathcal{M}(\gamma) := \{(r, p, A) \in \mathcal{J}^2 : r \leq -\gamma|p|\} \tag{5.1}$$

will be called the γ-*monotonicity cone subequation* and, given a real number $R \in (0, +\infty)$, the set

$$\mathcal{M}(R) := \{(r, p, A) \in \mathcal{J}^2 : A \geq \tfrac{|p|}{R} I\} \qquad (5.2)$$

will be called the *R-monotonicity cone subequation.*

Besides helping to complete a fundamental family, these new building blocks are interesting in their own right. They are all closed convex cones with vertex at the origin. It is also clear that they satisfy the subequation constraint conditions so that each is a monotonicity cone subequation. We will see that the cone $\mathcal{M}(\gamma)$ arises naturally in equations with strict monotonicity in r and a Lipschitz bound in p (see Theorem 12.8). Similarly, $\mathcal{M}(R)$ arises in zero-order-free equations with some degree of strict ellipticity and a Lipschitz bound in p (see Theorem 12.11).

Additional monotonicity cone subequations are generated by taking intersections of the 5 basic cones $\mathcal{M}(\mathcal{N})$, $\mathcal{M}(\mathcal{P})$, $\mathcal{M}(\mathcal{D})$, $\mathcal{M}(\gamma)$, and $\mathcal{M}(R)$. This provides us with our list of 17 distinct monotonicity cone subequations.

Definition 5.2 (Our list of monotonicity cone subequations).

(I) First on the list are 4 of the 5 basic examples defined above:

  (1) $\mathcal{M}(\mathcal{N})$;
  (2) $\mathcal{M}(\mathcal{P})$;
  (3) $\mathcal{M}(\gamma)$ with $\gamma \in (0, +\infty)$;
  (4) $\mathcal{M}(R)$ with $\mathbb{R} \in (0, +\infty)$.

(II) Double intersections of those in (I) give rise to 4 new monotonicity cone subequations:

  (5) $\mathcal{M}(\gamma, R) := \mathcal{M}(\gamma) \cap \mathcal{M}(R) = \{(r, p, A) \in \mathcal{J}^2 : r \leq -\gamma|p| \text{ and } A \geq \tfrac{|p|}{R} I\}$;
  (6) $\mathcal{M}(\gamma, \mathcal{P}) := \mathcal{M}(\gamma) \cap \mathcal{M}(\mathcal{P}) = \{(r, p, A) \in \mathcal{J}^2 : r \leq -\gamma|p| \text{ and } A \geq 0\}$;
  (7) $\mathcal{M}(\mathcal{N}, R) := \mathcal{M}(\mathcal{N}) \cap \mathcal{M}(R) = \{(r, p, A) \in \mathcal{J}^2 : r \leq 0 \text{ and } A \geq \tfrac{|p|}{R} I\}$;
  (8) $\mathcal{M}(\mathcal{N}, \mathcal{P}) := \mathcal{M}(\mathcal{N}) \cap \mathcal{M}(\mathcal{P}) = \{(r, p, A) \in \mathcal{J}^2 : r \leq 0 \text{ and } A \geq 0\}$.

(III) To complete the list, first add our last basic example:

  (9) $\mathcal{M}(\mathcal{D})$ with $\mathcal{D} \subsetneq \mathbb{R}^n$ a proper directional cone.

(IV) Intersecting such an $\mathcal{M}(\mathcal{D})$ with the first 8 examples completes the list:

  (10) $\mathcal{M}(\mathcal{N}, \mathcal{D}) := \mathcal{M}(\mathcal{N}) \cap \mathcal{M}(\mathcal{D}) = \{(r, p, A) \in \mathcal{J}^2 : r \leq 0 \text{ and } p \in \mathcal{D}\}$;
  (11) $\mathcal{M}(\mathcal{D}, \mathcal{P}) := \mathcal{M}(\mathcal{D}) \cap \mathcal{M}(\mathcal{P}) = \{(r, p, A) \in \mathcal{J}^2 : p \in \mathcal{D} \text{ and } A \geq 0\}$;
  (12) $\mathcal{M}(\gamma, \mathcal{D}) := \mathcal{M}(\gamma) \cap \mathcal{M}(\mathcal{D}) = \{(r, p, A) \in \mathcal{J}^2 : r \leq -\gamma|p| \text{ and } p \in \mathcal{D}\}$;
  (13) $\mathcal{M}(\mathcal{D}, R) := \mathcal{M}(\mathcal{D}) \cap \mathcal{M}(R) = \{(r, p, A) \in \mathcal{J}^2 : p \in \mathcal{D} \text{ and } A \geq \tfrac{|p|}{R} I\}$;
  (14) $\mathcal{M}(\gamma, \mathcal{D}, R) := \mathcal{M}(\gamma) \cap \mathcal{M}(\mathcal{D}) \cap \mathcal{M}(R)$, that is,

$$\mathcal{M}(\gamma, \mathcal{D}, R) = \{(r, p, A) \in \mathcal{J}^2 : r \leq -\gamma|p|, \ p \in \mathcal{D}, \text{ and } A \geq \tfrac{|p|}{R} I\};$$

  (15) $\mathcal{M}(\gamma, \mathcal{D}, \mathcal{P}) := \mathcal{M}(\gamma, \mathcal{P}) \cap \mathcal{M}(\mathcal{D}) = \mathcal{M}(\gamma) \cap \mathcal{M}(\mathcal{D}) \cap \mathcal{M}(\mathcal{P})$;
  (16) $\mathcal{M}(\mathcal{N}, \mathcal{D}, R) := \mathcal{M}(\mathcal{N}, R) \cap \mathcal{M}(\mathcal{D}) = \mathcal{M}(\mathcal{N}) \cap \mathcal{M}(\mathcal{D}) \cap \mathcal{M}(R)$;
  (17) $\mathcal{M}(\mathcal{N}, \mathcal{D}, \mathcal{P}) := \mathcal{M}(\mathcal{N}, \mathcal{P}) \cap \mathcal{M}(\mathcal{D}) = \mathcal{M}(\mathcal{N}) \cap \mathcal{M}(\mathcal{D}) \cap \mathcal{M}(\mathcal{P})$.

The cone $\mathcal{M}(\gamma, R)$ will be called the (γ, R)-*monotonicity cone subequation*. The same nomenclature will be used for the other cones above; for example, $\mathcal{M}(\gamma, \mathcal{D}, R)$ will be called the $(\gamma, \mathcal{D}, R)$-*monotonicity cone subequation*.

A few remarks are in order about this family of cones.

Remark 5.3. While taking all possible intersections of the basic 5 cones will produce more than 12 additional sets, many of the intersections are not distinct, since for each $\gamma, R \in (0, +\infty)$ one has

$$\mathcal{M}(\gamma) \subsetneq \mathcal{M}(\mathcal{N}) \quad \text{and} \quad \mathcal{M}(R) \subsetneq \mathcal{M}(\mathcal{P}) \quad \text{are proper subsets.} \tag{5.3}$$

Consequently, in (II), the double intersections $\mathcal{M}(\gamma) \cap \mathcal{M}(\mathcal{N}) = \mathcal{M}(\gamma)$ and $\mathcal{M}(R) \cap \mathcal{M}(\mathcal{P}) = \mathcal{M}(R)$ can be ignored, along with the triple intersections involving them and $\mathcal{M}(\mathcal{N}) \cap \mathcal{M}(\mathcal{P}) \cap \mathcal{M}(\gamma) \cap \mathcal{M}(R) = \mathcal{M}(\gamma, R)$.

Some of the cones have been seen before.

Remark 5.4 (Fundamental products). The elementary cones $\mathcal{M}(\mathcal{N})$, $\mathcal{M}(\mathcal{D})$, and $\mathcal{M}(\mathcal{P})$ are all examples of fundamental product monotonicity subequations in the sense of Definition 4.10, as are their intersections $\mathcal{M}(\mathcal{N}, \mathcal{P})$, $\mathcal{M}(\mathcal{N}, \mathcal{D})$, $\mathcal{M}(\mathcal{D}, \mathcal{P})$, and $\mathcal{M}(\mathcal{N}, \mathcal{D}, \mathcal{P})$.

Some of the cones have silent factors, which provide useful simplifications in the reductions of Chapter 10.

Remark 5.5 (Reduction by suppressing trivial factors). With the exception of $\mathcal{M}(\mathcal{N}, \mathcal{D}, \mathcal{P})$ in the case $\mathcal{D} \neq \mathbb{R}^n$, all of the cones in Remark 5.4 have a trivial factor $\mathbb{R}$, $\mathbb{R}^n$, or $\mathcal{S}(n)$. A particularly important case is the *negativity–convexity subequation* $\mathcal{M}(\mathcal{N}, \mathcal{P}) = \mathcal{N} \times \mathbb{R}^n \times \mathcal{P}$, which will de discussed at length in Section 10.2 on comparison for *gradient-free* subequations $\mathcal{F}$. In this case, one can *reduce* to the monotonicity cone $\mathcal{Q} := \mathcal{N} \times \mathcal{P} \subset \mathbb{R} \times \mathcal{S}(n)$.

The other reducible cones are $\mathcal{M}(\gamma)$, $\mathcal{M}(R)$, $\mathcal{M}(\gamma, \mathcal{D})$, and $\mathcal{M}(\mathcal{D}, R)$. If one eliminates the trivial factor, one can define *reduced cones* such as

$$\mathcal{M}'(\gamma, \mathcal{D}) := \{(r, p) \in \mathbb{R} \times \mathbb{R}^n : r \leq -\gamma|p| \text{ and } p \in \mathcal{D}\} \tag{5.4}$$

and

$$\mathcal{M}'(\mathcal{D}, R) := \{(p, A) \in \mathbb{R}^n \times \mathcal{S}(n) : p \in \mathcal{D} \text{ and } A \geq \tfrac{|p|}{R} I\}, \tag{5.5}$$

so that

$$\mathcal{M}(\gamma, \mathcal{D}) = \mathcal{M}'(\gamma, \mathcal{D}) \times \mathcal{S}(n) \quad \text{and} \quad \mathcal{M}(\mathcal{D}, R) = \mathbb{R} \times \mathcal{M}'(\mathcal{D}, R). \tag{5.6}$$

The cones in (5.5) will be used for zero-order-free subequations in Section 10.4. The other pair of reducible cones $\mathcal{M}(\gamma)$ and $\mathcal{M}(R)$ correspond to $\mathcal{D} = \mathbb{R}^n$ and we will write

$$\mathcal{M}'(\gamma) := \big\{(r,p) \in \mathbb{R} \times \mathbb{R}^n : r \le -\gamma|p|\big\} \tag{5.7}$$

and

$$\mathcal{M}'(R) := \big\{(p,A) \in \mathbb{R}^n \times \mathcal{S}(n) : A \ge \tfrac{|p|}{R} I\big\}. \tag{5.8}$$

The utility of $\mathcal{M}'(\gamma)$ and $\mathcal{M}'(R)$ was noted following Definition 5.1.

Finally, for future use, we record the following formulas for strictness and duality.

Remark 5.6. If $\mathcal{M} = \mathcal{M}(\gamma, \mathcal{D}, R)$ then

$$\operatorname{Int}\mathcal{M} = \big\{(r,p,A) \in \mathcal{J}^2 : r < -\gamma|p|, \ p \in \operatorname{Int}\mathcal{D}, \ \text{and} \ A > \tfrac{|p|}{R} I\big\}, \tag{5.9}$$

$$\widetilde{\mathcal{M}} = \big\{(r,p,A) \in \mathcal{J}^2 : \text{either} \ r \le \gamma|p|, \ \text{or} \ p \in \widetilde{\mathcal{D}}, \ \text{or} \ A + \tfrac{|p|}{R} I \in \widetilde{\mathcal{P}}\big\}, \tag{5.10}$$

$$\sim\!\widetilde{\mathcal{M}} = \big\{(r,p,A) \in \mathcal{J}^2 : r > \gamma|p|, \ -p \in \operatorname{Int}\mathcal{D}, \ \text{and} \ A < -\tfrac{|p|}{R} I\big\}. \tag{5.11}$$

Similar formulas for the other cones defined above are easily deduced from these.

5.2   NESTING, LIMIT CASES, AND SIMPLIFYING THE FAMILY OF CONES

In this section we look at the limiting cases of $\mathcal{M}(\gamma, \mathcal{D}, R)$-cones when the parameters γ and R in $(0, +\infty)$ tend to their limiting values. This analysis will allow us to express every element of the fundamental family in terms of a triple $(\gamma, \mathcal{D}, R)$ with $\gamma \in [0, +\infty)$, $\mathcal{D} \subseteq \mathbb{R}^N$ a proper directional cone or all of $\mathbb{R}^n$, and $R \in (0, +\infty]$. This simplification is carried out in Remark 5.9 below. With respect to the partial ordering by set inclusion, $\mathcal{M}(\gamma)$ and $\mathcal{M}(R)$ are nested families. It is easy to see from the definitions that

$$\mathcal{M}(\gamma) \text{ is decreasing in } \gamma \in (0, +\infty) \tag{5.12}$$

and

$$\mathcal{M}(R) \text{ is increasing in } R \in (0, +\infty). \tag{5.13}$$

Hence,

$$\mathcal{M}(\gamma, R) \text{ decreases as } \gamma \text{ increases and } R \text{ decreases}, \tag{5.14}$$

and $\mathcal{M}(\gamma, R)$ increases as γ decreases and R increases. Moreover, these monotonicity properties pass to intersections with $\mathcal{M}(\mathcal{D})$.

Proposition 5.7 (Decreasing limits). *For the family of* $(\gamma, \mathcal{D}, R)$-*cones defined in Definition 5.2(14), the decreasing limits*[1]

$$\lim_{\gamma \nearrow +\infty} \mathcal{M}(\gamma, \mathcal{D}, R), \quad \lim_{R \searrow 0} \mathcal{M}(\gamma, \mathcal{D}, R), \quad and \quad \lim_{\substack{\gamma \nearrow +\infty \\ R \searrow 0}} \mathcal{M}(\gamma, \mathcal{D}, R) \qquad (5.15)$$

are all equal to the minimal monotonicity set $\mathcal{M}_0 = \mathcal{N} \times \{0\} \times \mathcal{P}$.

Proof. First, note that

$$\lim_{\gamma \nearrow +\infty} \mathcal{M}(\gamma) = \bigcap_{\gamma > 0} \mathcal{M}(\gamma) = \mathcal{N} \times \{0\} \times \mathcal{S}(n)$$

and

$$\lim_{R \searrow 0} \mathcal{M}(R) = \bigcap_{R > 0} \mathcal{M}(R) = \mathbb{R} \times \{0\} \times \mathcal{P}$$

since

$$\mathcal{M}(\gamma) = \left\{ (r, p, A) \in \mathcal{J}^2 : |p| \leq -\tfrac{r}{\gamma} \right\} \quad \text{and} \quad \mathcal{M}(R) = \left\{ (r, P, A) \in \mathcal{J}^2 : |p| I \leq RA \right\}.$$

It follows easily that each of the three decreasing limits

$$\lim_{\gamma \nearrow +\infty} \mathcal{M}(\gamma, R), \quad \lim_{R \searrow 0} \mathcal{M}(\gamma, R), \quad and \quad \lim_{\substack{\gamma \nearrow +\infty \\ R \searrow 0}} \mathcal{M}(\gamma, R) \qquad (5.16)$$

decreases to $\mathcal{M}_0$. The role of $\mathcal{D}$ (a proper cone in $\mathbb{R}^n$) is innocuous and (5.15) follows immediately from (5.16) by intersecting with $\mathcal{M}(\mathcal{D})$. $\qquad \square$

For the increasing limits, first note that

$$\lim_{\gamma \searrow 0} \mathcal{M}(\gamma) = \bigcup_{\gamma > 0} \mathcal{M}(\gamma) = \mathcal{M}(\mathcal{N}) \setminus (\{0\} \times (\mathbb{R}^n \setminus \{0\}) \times \mathcal{S}(n)), \qquad (5.17)$$

which has closure $\mathcal{M}(\mathcal{N})$, and

$$\lim_{R \nearrow +\infty} \mathcal{M}(R) = \bigcup_{R > 0} \mathcal{M}(R) = \mathcal{M}(\mathcal{P}) \setminus (\mathbb{R} \times (\mathbb{R}^n \setminus \{0\}) \times \{0\}), \qquad (5.18)$$

which has closure $\mathcal{M}(\mathcal{P})$.

From these two facts we leave it to the reader to prove the following result.

---

[1]From the monotonicity of (5.12), one means, of course, the intersection over $\gamma \in (0, +\infty)$ of $\mathcal{M}(\gamma, \mathcal{D}, R)$ for the first decreasing limit and similar intersections for the last two.

Proposition 5.8 (Increasing limits). *For the family of $(\gamma, \mathcal{D}, R)$-cones defined in Definition 5.2(14), the following hold:*

(a) $\mathcal{M}'(\gamma, \mathcal{D}) \times \mathcal{P}$ *is the closure of the increasing limit* $\lim_{R \nearrow +\infty} \mathcal{M}(\gamma, \mathcal{D}, R)$;
(b) $\mathcal{N} \times \mathcal{M}'(\mathcal{D}, R)$ *is the closure of the increasing limit* $\lim_{\gamma \searrow 0} \mathcal{M}(\gamma, \mathcal{D}, R)$;
(c) $\mathcal{N} \times \mathcal{D} \times \mathcal{P}$ *is the closure of the increasing limit* $\lim_{\substack{\gamma \searrow 0 \\ R \nearrow +\infty}} \mathcal{M}(\gamma, \mathcal{D}, R)$.

Remark 5.9 (Simplifying the list of cones). In light of (5.17) and (5.18), it is natural to extend the parameters to allow $\gamma = 0$ and $R = +\infty$ and to define

$$\mathcal{M}(\gamma = 0) := \mathcal{M}(\mathcal{N}) \quad \text{and} \quad \mathcal{M}(R = +\infty) := \mathcal{M}(\mathcal{P}). \tag{5.19}$$

Also note that in the definition of $\mathcal{M}(\gamma)$, if one simply sets $\gamma = 0$ one obtains $\mathcal{M}(\mathcal{N})$. Similarly, one obtains $\mathcal{M}(\mathcal{P})$ by setting $R = +\infty$ in the definition of $\mathcal{M}(R)$. With such a choice, the five basic cones can be simplified to three:

$$\mathcal{M}(\gamma) \text{ with } \gamma \in [0, +\infty), \quad \mathcal{M}(R) \text{ with } R \in (0, +\infty], \quad \text{and} \quad \mathcal{M}(\mathcal{D}).$$

The remaining cones take the forms

$$\mathcal{M}(\gamma, R), \quad \mathcal{M}(\gamma, \mathcal{D}), \quad \mathcal{M}(\mathcal{D}, R), \quad \text{and} \quad \mathcal{M}(\gamma, \mathcal{D}, R). \tag{5.20}$$

Moreover, when $\mathcal{D} = \mathbb{R}^n$, one has $\mathcal{M}(\mathcal{D}) = \mathcal{J}^2$ and hence $\mathcal{M}(\gamma, \mathbb{R}^n, R) = \mathcal{M}(\gamma, R)$ and so on. Adopting these conventions/definitions, each of the 17 cones in Definition 5.2 is, in fact, a $(\gamma, \mathcal{D}, R)$-cone with $\gamma \in [0, +\infty)$, $R \in (0, +\infty]$, and $\mathcal{D} \subset \mathbb{R}^n$ either a proper directional cone or all of $\mathbb{R}^n$.

5.3 THE FUNDAMENTAL NATURE OF THE FAMILY OF MONOTONICITY CONES

Our family of monotonicity cones $\mathcal{M}(\gamma, \mathcal{D}, R)$ is "fundamental" in the following sense.

Theorem 5.10 (Fundamental family theorem). *If $\mathcal{F}$ is a subequation which is $\mathcal{M}$-monotone for some monotonicity cone subequation $\mathcal{M}$, then $\mathcal{F}$ is $\mathcal{M}(\gamma, \mathcal{D}, R)$-monotone for some $\gamma, R \in (0, +\infty)$ and some directional cone $\mathcal{D}$.*

Proof. It suffices to find $\mathcal{M}(\gamma, \mathcal{D}, R) \subset \mathcal{M}$. This follows from the next two lemmas. $\qquad\square$

Lemma 5.11. *Given a monotonicity cone subequation $\mathcal{M}$, if there exist $\varepsilon > 0$ and a directional cone $\mathcal{D} \subset \mathbb{R}^n$ such that $\{-1\} \times (\mathcal{D} \cap B_\varepsilon(0)) \times \{I\} \subset \mathcal{M}$, then* *the $(\gamma, \mathcal{D}, R)$-monotonicity cone with $R < \varepsilon$ and $\gamma := 1/R$ satisfies*

$$\mathcal{M}(\gamma, \mathcal{D}, R) \subset \mathcal{M}.$$

Proof. Suppose that $(r, p, A) \in \mathcal{M}(\gamma, \mathcal{D}, R)$. If $p = 0$, then $r \leq -\gamma |p| = 0$ and $A \geq \frac{|p|}{R} I = 0$, so that $(r, p, A) \in \mathcal{M}_0 \subset \mathcal{M}$.

Suppose now that $p \neq 0$. It suffices to show that

$$(r, p, A) = \frac{|p|}{R}\left[\left(-1, \frac{R}{|p|}p, I\right) + (-s, 0, P)\right] \tag{5.21}$$

with $s \geq 0$, $P \geq 0$, and $\frac{R}{|p|}p \in \mathcal{D} \cap B_\varepsilon(0)$, because then the facts

(i) $(-1, \frac{R}{|p|}p, I) \in \mathcal{M}$ by hypothesis;
(ii) $\mathcal{M}$ is $\mathcal{N} \times \{0\} \times \mathcal{P}$-monotone;
(iii) $\mathcal{M}$ is a cone;

combined with (5.21) prove that $(r, p, A) \in \mathcal{M}$. To see that (5.21) is true, note that this equality defines

$$s = -\frac{rR}{|p|} \quad \text{and} \quad P = A - \frac{|p|}{R}I.$$

Now, by the definition of $(r, p, A) \in \mathcal{M}(1/R, \mathcal{D}, R)$, one has $r \leq -\frac{|p|}{R}$ and $A \geq \frac{|p|}{R}I$. Hence, $P \geq 0$ and $\frac{(-r)R}{|p|} \geq 1$, so that $s \geq 0$ and $\frac{R}{|p|}p \in \mathcal{D} \cap B_\varepsilon(0)$ as $\mathcal{D}$ is a cone and $R < \varepsilon$. $\qquad\square$

Lemma 5.12. *Given a monotonicity cone subequation $\mathcal{M}$, there exist $\varepsilon > 0$ and a directional cone $\mathcal{D} \subset \mathbb{R}^n$ such that*

$$\{-1\} \times (\mathcal{D} \cap B_\varepsilon(0)) \times \{I\} \subset \mathcal{M}.$$

Proof. By the topological condition (T), $\operatorname{Int} \mathcal{M} \neq \emptyset$. Pick $(r, p, A) \in \operatorname{Int} \mathcal{M}$. By perturbing p we can assume that $p \neq 0$. Pick $\delta \in (0, |p|)$ small so that $\{r\} \times B_\delta(p) \times A \subset \operatorname{Int} \mathcal{M}$. For $t \geq t_0 > 0$ large, one has $P := tI - A \geq 0$ and $s := r + t \geq 0$. Hence,

$$(-t, q, tI) = (r, q, A) + (-s, 0, P) \in \operatorname{Int} \mathcal{M} \quad \forall q \in B_\delta(p), \ \forall t \geq t_0 > 0.$$

Since $\operatorname{Int} \mathcal{M}$ is a cone,

$$\{-1\} \times \frac{1}{t}B_\delta(p) \times \{I\} \subset \operatorname{Int} \mathcal{M} \quad \text{for each } t \geq t_0 > 0.$$

Take $\varepsilon := \delta/t_0$ and $\mathcal{D}$ the cone on $B_\delta(p)$. Then

$$\mathcal{D} \cap B_\varepsilon(0) \subset \bigcup_{t \geq t_0} \frac{1}{t} B_\delta(0) \cup \{0\},$$

which proves that $\{-1\} \times (\mathcal{D} \cap B_\varepsilon(0)) \times \{I\} \subset \mathcal{M}$. $\square$

It is important to note that the fundamental nature of the family will ensure the validity of the comparison principle locally (see Theorem 7.8).

Chapter Six

The Zero Maximum Principle for Dual Monotonicity Cones

In this chapter we examine the validity of the zero maximum principle (ZMP) for $\widetilde{\mathcal{M}}$-subharmonic functions if $\mathcal{M}$ is a monotonicity cone subequation. Its validity can be reduced to the existence of a global, regular, and strictly $\mathcal{M}$-subharmonic function. This function generates an approximation from above of the $\widetilde{\mathcal{M}}$-subharmonic function zero and has the advantage that the definitional comparison of Lemma 3.14 (see formula (6.4) below) applies since it is regular and strict (unlike zero).

Definition 6.1 (Strict approximator). Suppose that $\mathcal{M}$ is a convex cone subequation, that is, a convex cone with vertex at the origin which satisfies the subequation constraint conditions (P), (N), and (T). Given a domain $\Omega \subset\subset \mathbb{R}^n$, we say $\mathcal{M}$ *admits a strict approximator on* Ω if there exists ψ with

$$\psi \in C(\overline{\Omega}) \cap C^2(\Omega) \quad \text{and} \quad J_x^2 \psi \in \operatorname{Int} \mathcal{M} \text{ for each } x \in \Omega. \tag{6.1}$$

Theorem 6.2 (Zero maximum principle). *Suppose that $\mathcal{M}$ is a convex cone subequation that admits a strict approximator on Ω. Then the* (ZMP) *holds for $\widetilde{\mathcal{M}}$ on $\overline{\Omega}$, that is,*

$$z \leq 0 \text{ on } \partial\Omega \;\Rightarrow\; z \leq 0 \text{ on } \Omega \tag{6.2}$$

for all $z \in \operatorname{USC}(\overline{\Omega})$ which are $\widetilde{\mathcal{M}}$-subharmonic on Ω.

Notice that the (ZMP) is the comparison principle for $\widetilde{\mathcal{M}}$ in the case where $u = z$ and $w \equiv 0$, because by assumption $u := z$ is $\widetilde{\mathcal{M}}$-subharmonic and $w := 0$ is $\widetilde{\mathcal{M}}$-superharmonic since $J_x^2(-w) \equiv (0,0,0) \in \mathcal{M}$. The proof is an elementary consequence of the definitions (as is the proof of definitional comparison).

Proof. The dual $\widetilde{\mathcal{M}}$ also satisfies (P), (N), and (T) by property (7) of Proposition 3.2. Since $\widetilde{\mathcal{M}}$ has property (N), $z - m \in \widetilde{\mathcal{M}}(\Omega)$ for each $m \in (0, +\infty)$, as noted in Remark 2.12.

Since $z - m < 0$ on $\partial\Omega$ which is compact, one has

$$z - m + \varepsilon\psi \leq 0 \text{ on } \partial\Omega \quad \text{for each } \varepsilon > 0 \text{ sufficiently small.} \tag{6.3}$$

Now, since $z - m \in \widetilde{\mathcal{M}}(\Omega)$ and since $\varepsilon\psi \in C(\overline{\Omega}) \cap C^2(\Omega)$ is strictly $\mathcal{M}$-subharmonic on Ω (by coherence and $\mathcal{M}$ being a cone), one has

$$z - m + \varepsilon\psi \leq 0 \text{ on } \Omega \tag{6.4}$$

by the definitional comparison of Lemma 3.14 with $\mathcal{F} = \widetilde{\mathcal{M}}$ and $\widetilde{\mathcal{F}} = \widetilde{\widetilde{\mathcal{M}}} = \mathcal{M}$. Taking the limit in (6.4) as $m \searrow 0$ and $\varepsilon \searrow 0$ gives $z \leq 0$ on Ω. $\qquad\square$

As a corollary to this general theorem, we obtain the (ZMP) for each $(\gamma, \mathcal{D}, R)$-monotonicity cone as in Definition 5.2 (see also Remark 5.9), with a restriction on the size of the domain if R is finite. The following result was originally given in [50, Theorem B.2].

Theorem 6.3. *Let $\mathcal{M}$ be a $(\gamma, \mathcal{D}, R)$-monotonicity cone subequation. Given $z \in \mathrm{USC}(\overline{\Omega})$ which is $\widetilde{\mathcal{M}}$-subharmonic on Ω, one has*

$$z \leq 0 \text{ on } \partial\Omega \implies z \leq 0 \text{ on } \Omega \tag{ZMP}$$

as follows:

Case $R = +\infty$: *for arbitrary $\Omega \subset\subset \mathbb{R}^n$. This case includes $\mathcal{M}(\gamma, \mathcal{P}) = \mathcal{M}'(\gamma) \times \mathcal{P}$ with $\gamma \in [0, +\infty)$, where the case $\gamma = 0$ is $\mathcal{N} \times \mathbb{R}^n \times \mathcal{P} = \mathcal{M}(\mathcal{N}, \mathcal{P})$ and the case*

$$\mathcal{M}(\gamma, \mathcal{D}, \mathcal{P}) := \mathcal{M}(\gamma) \cap \mathcal{M}(\mathcal{D}) \cap \mathcal{M}(\mathcal{P}) := \mathcal{M}'(\gamma, \mathcal{D}) \times \mathcal{P},$$

and hence any of the larger monotonicity cone subequations, namely

$$\mathcal{M}(\gamma), \quad \mathcal{M}(\mathcal{D}), \quad \mathcal{M}(\mathcal{P}), \quad \mathcal{M}(\gamma, \mathcal{D}), \quad \text{and } \mathcal{M}(\mathcal{D}, \mathcal{P}).$$

Case R finite: *for domains Ω which are contained in a translate of the truncated cone $\mathcal{D}_R := \mathcal{D} \cap B_R(0)$, that is,*

$$\Omega \subset (y + \mathcal{D}) \cap B_R(y) \quad \text{for some } y \in \mathbb{R}^n. \tag{6.5}$$

This case includes $\mathcal{M}(\gamma, \mathcal{D}, R) := \mathcal{M}(\gamma) \cap \mathcal{M}(\mathcal{D}) \cap \mathcal{M}(R)$ with R finite and hence any of the larger monotonicity cone subequations, namely

$$\mathcal{M}(R), \quad \mathcal{M}(\gamma, R), \quad \text{and } \mathcal{M}(\mathcal{D}, R), \quad \text{with } R \text{ finite.}$$

Proof. Since $\mathcal{M}$ is a monotonicity cone subequation, by Theorem 6.2 it suffices to show that Ω admits a strict approximator ψ. It can be constructed as a quadratic polynomial

$$\psi(x) := -c + \tfrac{1}{2}|x - y|^2,$$

with $c > 0$ and $y \in \mathbb{R}^n$ chosen to ensure that

$$J_x^2 \psi = (-c + \tfrac{1}{2}|x - y|^2, (x - y), I) \in \operatorname{Int} \mathcal{M} \quad \text{for every } x \in \Omega. \qquad (6.6)$$

Using the definition of the interior of $\mathcal{M}$ as given in (5.9), condition (6.6) requires that

$$RI > |x - y|I \ \text{ for every } x \in \Omega, \quad \text{that is, } \Omega \subset B_R(y); \qquad (6.7)$$
$$(x - y) \in \operatorname{Int} \mathcal{D} \ \text{ for every } x \in \Omega, \quad \text{that is, } \Omega \subset y + \operatorname{Int} \mathcal{D}; \qquad (6.8)$$

and finally that

$$-c + \tfrac{1}{2}|x - y|^2 < -\gamma |x - y| \quad \text{for every } x \in \Omega. \qquad (6.9)$$

Hypothesis (6.5) is equivalent to (6.7) and (6.8). By (6.7), choosing $c > \tfrac{1}{2}R^2 + \gamma R$ ensures (6.9), which completes the proof in the case $R < +\infty$.

In the case $R = +\infty$, condition (6.7) is automatic for each $y \in \mathbb{R}^n$ and since Ω is bounded, one can always pick $y \in \mathbb{R}^n$ such that (6.8) holds. Finally, choose any $R < +\infty$ with $\Omega \subset B_R(y)$ and choose $c > \tfrac{1}{2}R^2 + \gamma R$ to ensure (6.9). $\qquad \square$

We remark that Theorem 6.3 applies to all of the cones $\mathcal{M}$ in Definition 5.2. The cones in parts (I) and (II) correspond to the special case $\mathcal{D} = \mathbb{R}^n$. It is important to note that there are a priori restrictions on the domain when R is finite which can be essential, as will be shown in Proposition 6.5 below, whose proof applies the following important fact concerning reduced subequations to the dual $\widetilde{\mathcal{M}}$ of a reduced monotonicity cone subequation $\mathcal{M}$. As is standard in differential topology, *reduced* means that the jet variable $r \in \mathbb{R}$ is silent. As will be discussed in Chapter 10, this is equivalent to the following monotonicity property that strengthens property (N):

$$(r, p, A) \in \mathcal{M} \ \Rightarrow \ (r + s, p, A) \in \mathcal{M} \text{ for every } s \in \mathbb{R}. \qquad (6.10)$$

Note that $\mathcal{M}$ is reduced if and only if its dual $\widetilde{\mathcal{M}}$ is reduced.

Lemma 6.4. *Suppose that $\mathcal{G}$ is a reduced subequation in the sense of* (6.10). *Then the* (ZMP) *holds for $\mathcal{G}$ on $\overline{\Omega}$ if and only if the maximum principle* (MP) *holds for $\mathcal{G}$ on $\overline{\Omega}$, that is,*

$$\sup_{\Omega} u \leq \sup_{\partial \Omega} u \qquad \text{(MP)}$$

for each $u \in \operatorname{USC}(\overline{\Omega})$ which is $\mathcal{G}$-subharmonic on Ω.

Proof. The proof that the (MP) implies the (ZMP) on $\overline{\Omega}$ is immediate since if $z \in \operatorname{USC}(\overline{\Omega})$ is $\mathcal{G}$-subharmonic in Ω with $z \leq 0$ on $\partial \Omega$, the (MP) applied to $u = z$

gives $z \leq 0$ on Ω, as desired. Conversely, suppose that the (ZMP) holds and take any $u \in \mathrm{USC}(\overline{\Omega})$ which is $\mathcal{G}$-subharmonic in Ω. The function $z \in \mathrm{USC}(\overline{\Omega})$ defined by $z := u - \sup_{\partial\Omega} u$ satisfies $z \leq 0$ on $\partial\Omega$ and is $\mathcal{G}$-subharmonic in Ω (since $\mathcal{G}$ is a reduced subequation). Hence, $z \leq 0$ on Ω by the (ZMP). Note that u is $\mathcal{G}$-subharmonic if and only if $u - c$ is $\mathcal{G}$-subharmonic for any constant $c \in \mathbb{R}$. Now the result follows easily. $\qquad\square$

Proposition 6.5 (Failure of the (ZMP) for $\widetilde{\mathcal{M}}(R)$ on large balls). *In $\mathbb{R}^n$ with $n \geq 2$, consider the reduced (convex) monotonicity cone subequation*

$$\mathcal{M}(R) = \{(r, p, A) \in \mathcal{J}^2 : A \geq \tfrac{|p|}{R} I\} \quad \text{with } R \in (0, +\infty). \tag{6.11}$$

Then the (ZMP) for $\widetilde{M}(R)$ fails on $\overline{\Omega}$ with $\Omega = B_{R'}(0)$ for each $R' > R$.

Proof. Since $\widetilde{\mathcal{M}}(R)$ is a reduced subequation, the (ZMP) for $\widetilde{\mathcal{M}}(R)$ holds if and only if the (MP) holds for $\widetilde{\mathcal{M}}(R)$ (see Lemma 6.4). We exhibit a radial counterexample to the (MP) for $\widetilde{\mathcal{M}}(R)$ on $B_{R'}(0)$ with $R' > R$ using the radial calculations as recorded in Remark 3.17. Consider

$$u(x) := \psi(|x|) \quad \text{with } \psi(t) := t - \frac{t^2}{2R}. \tag{6.12}$$

Note that

$$\psi'(t) = \frac{R - t}{R} \quad \text{and} \quad \psi''(t) = -\frac{1}{R}, \tag{6.13}$$

and ψ has its only critical point in $t = R$ with global maximum value $R/2$. Hence,

$$u(x) := |x| - \frac{1}{R}\frac{|x|^2}{2} \text{ has its global maximum value on the sphere } |x| = R, \tag{6.14}$$

and hence *fails* to satisfy the (MP) on $\overline{\Omega}$ for any ball $\Omega = B_{R'}(0)$ with radius $R' > R$.

It remains only to show that u is $\widetilde{\mathcal{M}}(R)$-subharmonic on $\mathbb{R}^n$. It is easy to see that (use (5.10) with $\gamma = 0$ and $\mathcal{D} = \mathbb{R}^n$)

$$\widetilde{\mathcal{M}}(R) = \{(r, p, A) \in \mathcal{J}^2 : \lambda_{\max}(A) + \tfrac{|p|}{R} \geq 0\}. \tag{6.15}$$

The function $|x|$ does not have any upper test functions at $x = 0$, so neither does $|x|$ minus the quadratic $\frac{|x|^2}{2R}$. For $x \neq 0$, where u is smooth, we show that its 2-jet satisfies $J_x^2 u \in \widetilde{\mathcal{M}}(R)$ by using the radial calculus. For $t = |x| \neq 0$, using the radial formula (3.24) together with (6.13), we have

$$p := Du(x) = \psi'(|x|)\frac{x}{|x|} = \frac{R - |x|}{R}\frac{x}{|x|} \qquad (6.16)$$

and

$$A := D^2 u(x) = \frac{\psi'(|x|)}{|x|}P_{x^\perp} + \psi''(|x|)P_x = \left(-\frac{1}{R} + \frac{1}{|x|}\right)P_{x^\perp} - \frac{1}{R}P_x. \qquad (6.17)$$

Hence, for $n \geq 2$ we have $\lambda_{\max}(A) = -\frac{1}{R} + \frac{1}{|x|}$. In particular, if $0 < |x| \leq R$, $\lambda_{\max}(A) > 0$, and hence $\lambda_{\max}(A) + \frac{|p|}{R} > 0$. On the other hand, if $R < |x|$ then by (6.16) and (6.17) we have

$$\frac{|p|}{R} = \frac{|x| - R}{R} > 0 \quad \text{and} \quad \lambda_{\max}(A) = -\frac{(|x| - R)}{R|x|},$$

so that

$$\lambda_{\max}(A) + \frac{|p|}{R} = \frac{|x| - R}{R}\left[\frac{1}{R} - \frac{1}{|x|}\right] > 0. \qquad \square$$

Chapter Seven

Comparison Principles for Potential Theories with Sufficient Monotonicity

In this chapter we examine the central question of the book, which is the validity of comparison (C) for $\mathcal{F}$ on $\overline{\Omega}$:

$$u \leq w \text{ on } \partial\Omega \;\Rightarrow\; u \leq w \text{ on } \Omega, \tag{7.1}$$

or equivalently, the (ZMP) for the *comparison differences*

$$u - w \leq 0 \text{ on } \partial\Omega \;\Rightarrow\; u - w \leq 0 \text{ on } \Omega, \tag{7.2}$$

if $u \in \mathrm{USC}(\overline{\Omega})$ *and* $w \in \mathrm{LSC}(\overline{\Omega})$ *are* $\mathcal{F}$-*subharmonic and* $\mathcal{F}$-*superharmonic respectively on* Ω. As noted in the discussion of (1.42)–(1.43), if one uses Dirichlet duality and defines $v := -w$, comparison (C) is equivalent to the (ZMP) for sums (ZMP for sums) on $\overline{\Omega}$:

$$u + v \leq 0 \text{ on } \partial\Omega \;\Rightarrow\; u + v \leq 0 \text{ on } \Omega \tag{7.3}$$

if u *and* $v \in \mathrm{USC}(\overline{\Omega})$ *are* $\mathcal{F}$-*subharmonic and* $\widetilde{\mathcal{F}}$-*subharmonic respectively on* Ω. This second form (7.3) is the one which will be proved. Moreover, since $\widetilde{\widetilde{\mathcal{F}}} = \mathcal{F}$, version (7.3) of comparison immediately implies the symmetry

$$\text{comparison for } \mathcal{F} \text{ on } \overline{\Omega} \;\Leftrightarrow\; \text{comparison for } \widetilde{\mathcal{F}} \text{ on } \overline{\Omega}. \tag{7.4}$$

Our method is dependent on being able to find a subequation $\mathcal{H}$ with two properties:

$$\mathcal{F}(X) + \widetilde{\mathcal{F}}(X) \subset \mathcal{H}(X) \quad \text{for every open set } X \subset \mathbb{R}^n, \tag{7.5}$$

and $\mathcal{H}(X)$ satisfying the (ZMP), that is,

$$h \leq 0 \text{ on } \partial\Omega \;\Rightarrow\; h \leq 0 \text{ on } \Omega \;\forall\, \Omega \subset\subset X, \; h \in \mathrm{USC}(\overline{\Omega}) \cap \mathcal{H}(\Omega). \tag{7.6}$$

The first step is to find $\mathcal{H}$ satisfying (7.5) and the second step is to show that (7.6) holds. We will discover $\mathcal{H}$ infinitesimally, which reduces to monotonicity by using duality. At the infinitesimal (2-jet) level, $\mathcal{H}$ must be the dual $\widetilde{\mathcal{M}}$ of a monotonicity set $\mathcal{M}$ for $\mathcal{F}$. This is done in Lemma 7.3 below, but first we prove that a *subharmonic addition* such as (7.5) is implied by its infinitesimal version, *jet addition*.

Theorem 7.1 (Subharmonic addition theorem). *For arbitrary subequation constraint sets $\mathcal{F}$, $\mathcal{G}$, and $\mathcal{H}$ of $\mathcal{J}^2$,*

$$\text{(jet addition)} \quad \mathcal{F} + \mathcal{G} \subset \mathcal{H} \tag{7.7}$$

implies

$$\text{(subharmonic addition)} \quad \mathcal{F}(X) + \mathcal{G}(X) \subset \mathcal{H}(X) \tag{7.8}$$

for the subharmonics on each open set $X \subset \mathbb{R}^n$.

We include the complete proof of this constant-coefficient result for the reader's convenience.

Proof. Given $u \in \mathcal{F}(X)$ and $v \in \mathcal{G}(X)$, it suffices to show that about each $x_0 \in X$ there is an open ball $B = B_\rho(x_0) \subset\subset \Omega$ such that $u + v \in \mathcal{H}(B)$. Since $u, v \in \text{USC}(X)$, they are bounded from above on any compact subset of X. Both u and v can be written as a decreasing limit of quasi-convex *sup convolution approximations* if u and v are also locally bounded from below.

To this end, by shrinking B if necessary, since $\mathcal{F}$ and $\mathcal{G}$ satisfy conditions (T) and (P), about each $x_0 \in X$ one can find quadratic functions φ, ψ which are $\mathcal{F}$-, $\mathcal{G}$-subharmonic and bounded on a common $B = B_\rho(x_0)$ (see Remark 2.13). The sequences of functions

$$u_m := \max\{u, \varphi - m\} \in \mathcal{F}(B) \quad \text{and} \quad v_m := \max\{u, \psi - m\} \in \mathcal{G}(B), \quad m \in \mathbb{N} \tag{7.9}$$

are bounded from above and below on B. The $\mathcal{F}$-, $\mathcal{G}$-subharmonicity claims use the maximum property (B) of Proposition D.1 and the negativity property (N) of $\mathcal{F}$, $\mathcal{G}$ applied to φ, ψ which are C^2.

Using these truncating approximations, in the proof of $u + v \in \mathcal{H}(B)$ we may assume that u, v are bounded on B, that is, there exists $N > 0$ such that

$$|u(x)|, |v(x)| \le N \quad \text{for all } x \in B. \tag{7.10}$$

Indeed, if the sum of the truncations in (7.9) satisfies $u_m + v_m \in \mathcal{H}(B)$ for each $m \in \mathbb{N}$, the decreasing sequence property (E) of Proposition D.1 shows that the limit satisfies $u + v \in \mathcal{H}(B)$, as desired.

Now, assuming (7.10), one passes to the sup convolutions

$$u^\varepsilon(x) := \sup_{y \in B}\left\{u(y) - \frac{1}{\varepsilon}|x - y|^2\right\}, \quad x \in B, \ \varepsilon > 0, \tag{7.11}$$

and similarly for v^ε. One has that u^ε, v^ε are $2/\varepsilon$-quasi-convex and decrease to u, v (where one uses that u, v are bounded below for the limit statement and hence the need for the truncation (7.9)). Moreover, one has that

$$u^\varepsilon \in \mathcal{F}(B_\delta) \quad \text{and} \quad v^\varepsilon \in \mathcal{G}(B_\delta), \tag{7.12}$$

where $B_\delta := \{x \in B : \text{dist}(x, \partial B) > \delta\}$ and $\delta = \sqrt{2\varepsilon N}$. One uses the translation property (D) and the families locally bounded above property (F) of Proposition D.1.

By Alexandroff's theorem, (7.12), and the jet addition hypothesis ($\mathcal{F} + \mathcal{G} \subset \mathcal{H}$), one has that the quasi-convex u^ε, v^ε satisfy

$$J_x^2(u^\varepsilon + v^\varepsilon) \in \mathcal{H} \quad \text{for almost every } x \in B_\delta. \tag{7.13}$$

For quasi-convex functions, statement (7.13) yields $u^\varepsilon + v^\varepsilon \in \mathcal{H}(B_\delta)$ by the almost everywhere theorem of Lemma 2.10. The desired conclusion follows from the decreasing sequence property (E) of Proposition D.1 by considering the limit along a sequence corresponding to $\varepsilon = \varepsilon_j \to 0^+$.   □

Remark 7.2 (Subharmonic addition for locally quasi-convex functions). For locally quasi-convex functions, Theorem 7.1 extends from constant-coefficient subequations to arbitrary subequations, and hence from any open set X in Euclidean space to a manifold X. Namely,

$$\mathcal{F} + \mathcal{G} \subset \mathcal{H} \;\Rightarrow\; u + v \in \mathcal{H}(X) \;\, \forall u \in \mathcal{F}(X), \; v \in \mathcal{G}(X) \text{ locally quasi-convex.}$$

This is immediate from the proof of Theorem 7.1 above, since the sup convolution step is unnecessary if u and v are assumed to be locally quasi-convex, and the other steps do not use translation invariance.

In the special case $\mathcal{G} := \widetilde{\mathcal{F}}$, the subharmonic addition theorem concludes the desired subharmonic addition (7.5) stating that

$$\text{if } \mathcal{F} + \widetilde{\mathcal{F}} \subset \mathcal{H} \text{ then } \mathcal{F}(X) + \widetilde{\mathcal{F}}(X) \subset \mathcal{H}(X). \tag{7.14}$$

Next, using duality, we reduce the jet addition hypothesis $\mathcal{F} + \widetilde{\mathcal{F}} \subset \mathcal{H}$ to a monotonicity hypothesis $\mathcal{F} + \widetilde{\mathcal{H}} \subset \mathcal{F}$ on $\mathcal{F}$. This is a key step in the basic method of this monograph.

Lemma 7.3 (Jet addition, duality, and monotonicity). *For any two subequation constraint sets $\mathcal{F}, \mathcal{H} \subset \mathcal{J}^2$, one has*

$$\mathcal{F} + \widetilde{\mathcal{F}} \subset \mathcal{H} \;\Leftrightarrow\; \mathcal{F} + \widetilde{\mathcal{H}} \subset \mathcal{F}. \tag{7.15}$$

Proof. One sees that for $J = (r, p, A) \in \mathcal{J}^2$ one has

$$J + \widetilde{\mathcal{F}} \subset \mathcal{H} \iff J + \widetilde{\mathcal{H}} \subset \mathcal{F} \tag{7.16}$$

since $\widetilde{\mathcal{F} - J} = \widetilde{\mathcal{F}} + J \subset \mathcal{H} \iff \widetilde{\mathcal{H}} \subset \mathcal{F} - J \iff J + \widetilde{\mathcal{H}} \subset \mathcal{F}$, by the elementary properties (1) and (3) of the Dirichlet dual in Proposition 3.2. Taking all $J \in \mathcal{F}$ in (7.16) yields the lemma. $\square$

Consequently, finding a subequation $\mathcal{H}$ with the desired jet addition property $\mathcal{F} + \widetilde{\mathcal{F}} \subset \mathcal{H}$ requires that $\mathcal{H}$ equals the dual $\widetilde{\mathcal{M}}$ of a monotonicity subequation $\mathcal{M}$ for $\mathcal{F}$, that is, satisfying $\mathcal{F} + \mathcal{M} \subset \mathcal{F}$. We can summarize as follows.

Theorem 7.4 (Subharmonic addition, duality, and monotonicity). *Suppose that $\mathcal{M} \subset \mathcal{J}^2$ is a monotonicity cone subequation and that $\mathcal{F} \subset \mathcal{J}^2$ is an $\mathcal{M}$-monotone subequation constraint set. Then, for every open set $X \subset \mathbb{R}^n$, one has*

$$\mathcal{F}(X) + \widetilde{\mathcal{F}}(X) \subset \widetilde{\mathcal{M}}(X). \tag{7.17}$$

Combining Theorem 7.4 with Theorem 6.2 yields our general method for proving comparison.

Theorem 7.5 (General comparison theorem). *Suppose that a subequation $\mathcal{F} \subset \mathcal{J}^2$ is $\mathcal{M}$-monotone for some convex cone subequation $\mathcal{M}$. If $\mathcal{M}$ admits a strict approximator ψ on Ω (in the sense of Definition 6.1), then comparison (C) holds for $\mathcal{F}$ on $\overline{\Omega}$.*

Proof. Suppose that u and $v \in \mathrm{USC}(\overline{\Omega})$ are $\mathcal{F}$-subharmonic and $\widetilde{\widetilde{\mathcal{F}}}$-subharmonic respectively on Ω. Taking $X = \Omega$ in Theorem 7.4, we have $z := u + v \in \widetilde{\mathcal{M}}(\Omega)$, and hence $z \in \mathrm{USC}(\overline{\Omega}) \cap \widetilde{\mathcal{M}}(\Omega)$. By Theorem 6.2, since $\mathcal{M}$ has a strict approximator on Ω, such a z satisfies the (ZMP), that is,

$$u + v \leq 0 \text{ on } \partial\Omega \implies u + v \leq 0 \text{ on } \Omega. \tag{7.18}$$

This is precisely (7.3), which as noted above is one way of formulating the comparison principle (C). $\square$

We are now ready for the main result.

Theorem 7.6 (Fundamental family comparison theorem). *Suppose that $\mathcal{F} \subset \mathcal{J}^2$ is an $\mathcal{M}$-monotone subequation constraint set where $\mathcal{M} \subset \mathcal{J}^2$ is a monotonicity cone subequation. Given $u, v \in \mathrm{USC}(\overline{\Omega})$ which are $\mathcal{F}$-, $\widetilde{\mathcal{F}}$-subharmonic on Ω, one has*

$$u + v \leq 0 \text{ on } \partial\Omega \implies u + v \leq 0 \text{ on } \Omega \tag{C}$$

(1) *for each Ω contained in a translate of the truncated cone $\mathcal{D}_R := \mathcal{D} \cap B_R(0)$ if $\mathcal{M}$ contains one of the cones*

$$\mathcal{M}(\gamma, \mathcal{D}, R) := \mathcal{M}(\gamma) \cap \mathcal{M}(\mathcal{D}) \cap \mathcal{M}(R) \quad \text{with } R \text{ finite} \qquad (7.19)$$

and

(2) *for arbitrary $\Omega \subset\subset \mathbb{R}^n$ if $\mathcal{M}$ contains one of the cones*

$$\mathcal{M}(\gamma, \mathcal{D}, \mathcal{P}) := \mathcal{M}(\gamma) \cap \mathcal{M}(\mathcal{D}) \cap \mathcal{M}(\mathcal{P}) := \mathcal{M}'(\gamma, \mathcal{D}) \times \mathcal{P}. \qquad (7.20)$$

Moreover, by the fundamental family theorem (Theorem 5.10), every monotonicity cone subequation $\mathcal{M}$ contains a cone of type (7.19), so case (1) always holds. Finally, any cone of type (7.20) satisfies $\mathcal{M}(\gamma, \mathcal{D}, \mathcal{P}) \supset \mathcal{M}(\gamma, \mathcal{D}, R)$ for each R finite, and hence case (1) implies case (2).

Proof. Suppose that u and $v \in \mathrm{USC}(\overline{\Omega})$ are $\mathcal{F}$-subharmonic and $\widetilde{\mathcal{F}}$-subharmonic respectively on Ω. Again, by taking $X = \Omega$ in Theorem 7.4, we have $z := u + v \in \widetilde{\mathcal{M}}(\Omega)$ and hence $z \in \mathrm{USC}(\overline{\Omega}) \cap \mathcal{M}(\Omega)$. Therefore, it suffices to verify the (ZMP) for such z. In cases (1) and (2), this is exactly what Theorem 6.3 states in the cases R finite and $R = +\infty$, respectively. $\qquad\square$

The size of the domain Ω in Theorem 7.6(1) is sharp for the subequation $\mathcal{F} = \mathcal{M}(R)$ when R is finite.

Example 7.7. With $n \geq 2$ and R finite, comparison fails for $\mathcal{F} = \mathcal{M}(R)$ on $\overline{\Omega}$ with $\Omega = B_{R'}(0)$ for each $R' > R$. Indeed, as noted in (7.4), one has

$$\text{comparison for } \mathcal{F} = \mathcal{M}(R) \text{ on } \overline{\Omega} \iff \text{comparison for } \widetilde{\mathcal{F}} = \widetilde{\mathcal{M}}(R) \text{ on } \overline{\Omega}.$$

By Proposition 6.5 we know that the (ZMP) fails for $\widetilde{\mathcal{M}}(R)$ on $\overline{\Omega}$ with $\Omega = B_{R'}(0)$ for each $R' > R$, which completes the claim.

A larger family of subequations with maximal monotonicity $\mathcal{M}(R)$ and failure of comparison on balls of radius $R' > R$ will be presented in Example 9.1.

On the other hand, the fundamental nature of the family of $\mathcal{M}(\gamma, \mathcal{D}, R)$-cones gives rise to the local validity of the comparison principle for subequations with this minimal monotonicity.

Theorem 7.8 (Local comparison). *If $\mathcal{F}$ is a subequation which is $\mathcal{M}$-monotone for some monotonicity cone subequation $\mathcal{M}$, then the comparison principle holds locally on $\mathbb{R}^n$; in particular, there exists $\rho > 0$ which depends on $\mathcal{M}$ such that, for all domains $\Omega \subset B_\rho(x_0)$ with $x_0 \in \mathbb{R}^n$ arbitrary,*

$$u + v \leq 0 \text{ on } \partial\Omega \implies u + v \leq 0 \text{ on } \Omega \qquad (C)$$

for each pair $u \in \mathrm{USC}(\overline{\Omega}) \cap \mathcal{F}(\Omega)$ and $v \in \mathrm{USC}(\overline{\Omega}) \cap \widetilde{\mathcal{F}}(\Omega)$.

Proof. Since $\mathcal{F}$ is $\mathcal{M}$-monotone for a monotonicity cone subequation, by Theorem 5.10 there exists a cone $\mathcal{M}(\gamma, \mathcal{D}, R)$ in the fundamental family with $\mathcal{M}(\gamma, \mathcal{D}, R) \subset \mathcal{M}$ and hence $\mathcal{F}$ is $\mathcal{M}(\gamma, \mathcal{D}, R)$-monotone. Then, by Theorem 7.5, comparison holds on all domains Ω contained in a translate of the truncated cone $\mathcal{D}_R := \mathcal{D} \cap B_R(0)$. Clearly, there exists $\rho > 0$ such that $B_\rho(y_0) \subset \mathcal{D}_R$ for some $y_0 \in \mathcal{D}_R$ and hence $\Omega \subset B_\rho(x_0)$ is contained in a translation of $\mathcal{D}_R \supset B_\rho(y_0)$. $\quad\square$

We give one final comment in this chapter. In the proof of Theorem 7.4 (and hence for the proof of Theorem 7.6), one needed the local existence of bounded $\mathcal{F}$-, $\widetilde{\mathcal{F}}$-subharmonics on potentially small balls. However, one can also find subharmonics on potentially large balls in various ways. We record this observation for future reference.

Remark 7.9. If one knows the existence of particular 2-jets $(r_1, p_1, A_1) \in \mathcal{F}$, the construction of explicit bounded and smooth subharmonics simplifies considerably. For example, if there exists $(r_1, 0, 0) \in \mathcal{F}$, then any constant function $\varphi \equiv r_0$ with $r_0 \leq r_1$ will do by the negativity property (N). Moreover, if $(r_1, p_1, 0) \in \mathcal{F}$, then any affine function $\varphi(x) := r_0 + \langle p_0, x - x_0 \rangle$ will be $\mathcal{F}$-subharmonic on $B_\rho(x_0)$ if

$$p_0 := p_1 \quad \text{and} \quad r_0 - r_1 + \rho|p_0| \leq 0.$$

If one has neither of these two possibilities, about each $x_0 \in \Omega$ one can use the $\mathcal{M}$-monotonicity of $\mathcal{F}$ to construct quadratic polynomials

$$\varphi(x) := r_0 + \langle p_0, x - x_0 \rangle + \frac{\lambda_0}{2}|x - x_0|^2 \qquad (7.21)$$

with $r_0 < 0 < \lambda_0$, $p_0 \in \mathbb{R}^n$ chosen to ensure that $J_x^2 \varphi \in \mathcal{F}$ for each $x \in B_\rho(x_0) \subset \Omega$, for some $\rho > 0$. Starting from any $(r_1, p_1, A_1) \in \mathcal{F}$, one uses property (P) to show that $(r_1, p_1, \lambda_1 I) \in \mathcal{F}$ for λ_1 large enough. Using the $\mathcal{M}$-monotonicity it suffices to exhibit $(r_0, p_0, \lambda_0) \in \mathbb{R} \times \mathbb{R}^n \times \mathbb{R}$ and $\rho > 0$ such that, for $x \in B_\rho(x_0)$ and

$$J_x^2 \varphi = \big(\varphi(x), p_0 + \lambda_0(x - x_0), \lambda_0 I\big) := (r_1, p_1, \lambda_1 I) + \big(r(x), p(x), (\lambda_0 - \lambda_1)I\big), \qquad (7.22)$$

one has $(r(x), p(x), (\lambda_0 - \lambda_1)I) \in \mathcal{M}$, which requires

$$r(x) \leq -\gamma|p(x)|, \quad p(x) \in \mathcal{D}, \quad R(\lambda_0 - \lambda_1)I \geq |p(x)|I, \qquad (7.23)$$

where the last condition in (7.23) holds for every $\lambda_0 \geq \lambda_1$ in the case $R = +\infty$. The reader can verify easily that for a suitable radius ρ one can find (r_0, p_0, λ_0) for which (7.23) holds.

Chapter Eight

Comparison on Arbitrary Domains
by Additional Monotonicity

By Theorem 5.10, any subequation constraint set $\mathcal{F}$ which is $\mathcal{M}$-monotone for some monotonicity cone subequation $\mathcal{M}$ must have at least the monotonicity of one of the monotonicity cone subequations $\mathcal{M}(\gamma, \mathcal{D}, R) \subset \mathcal{M}$ belonging to our fundamental family. If $R = +\infty$, then comparison holds for $\mathcal{F}$ on arbitrary bounded domains by Theorem 7.6. If $R < +\infty$, then (again by Theorem 7.6) comparison holds for $\mathcal{F}$ on domains Ω for which a translate of Ω is contained in the truncated cone $\mathcal{D} \cap B_R(0)$. This result is sharp if the maximal monotonicity cone subequation $\mathcal{M}_{\mathcal{F}}$ for $\mathcal{F}$ is $\mathcal{M}(\gamma, \mathcal{D}, R)$ with R finite (see Proposition 9.2 for an example). However, this leaves room for improvement if $\mathcal{M}_{\mathcal{F}}$ is large enough, and this is the subject of the present chapter.

Comparison may still hold for all domains $\Omega \subset\subset \mathbb{R}^n$. We explore this possibility here, continuing with our monotonicity technique, looking for larger, not smaller, monotonicity cone subequations, and highlight two examples. These two examples contain

$$\mathcal{M}(R) = \left\{ (r, p, A) \in \mathcal{J}^2 : \lambda_{\min}(A) \geq \tfrac{|p|}{R} \right\}. \tag{8.1}$$

Definition 8.1. Fix $R \in (0, +\infty)$. Define

$$\mathcal{M}_R := \left\{ (r, p, A) \in \mathcal{J}^2 : A \geq 0 \text{ and } (\lambda_1(A) \cdots \lambda_n(A))^{1/n} \geq \tfrac{|p|}{R} \right\} \tag{8.2}$$

and

$$\mathcal{M}^R := \left\{ (r, p, A) \in \mathcal{J}^2 : A \geq 0 \text{ and } \langle Ae, e \rangle \geq \tfrac{|\langle p, e \rangle|}{R} \ \forall e \in \mathbb{R}^n \text{ with } |e| = 1 \right\}. \tag{8.3}$$

These variants of $\mathcal{M}(R)$ are indeed convex cone subequations and are all larger than $\mathcal{M}(R)$ in a precise sense.

Proposition 8.2. *For $R \in (0, +\infty)$ fixed,*

$$\mathcal{M}_R \text{ and } \mathcal{M}^R \text{ are convex cone subequations} \tag{8.4}$$

and

$$\mathcal{M}_R \text{ and } \mathcal{M}^R \text{ contain } \mathcal{M}(R') \iff R' \leq R. \tag{8.5}$$

Proof. $\mathcal{M}_R$ and $\mathcal{M}^R$ are convex cones since each is defined by an inequality of the form $h(p, A) \geq 0$, where h is a concave function on its domain. Property (N) is automatic as the variable r is silent in both cases, and property (P) follows since each $h(p, A)$ is increasing in A on its domain. Property (T) is satisfied since each $\mathcal{M}$ is a convex cone with Int $\mathcal{M}$ nonempty. □

Comparison always holds for all bounded domains for subequations $\mathcal{F}$ which are $\mathcal{M}$-monotone if $\mathcal{M}$ contains either $\mathcal{M}_R$ or $\mathcal{M}^R$.

Theorem 8.3. *Suppose that $\mathcal{F} \subset \mathcal{J}^2$ is a subequation which is $\mathcal{M}$-monotone. If $\mathcal{M}$ contains either $\mathcal{M}_R$ or $\mathcal{M}^R$ for some R, then comparison holds for $\mathcal{F}$ on all bounded domains $\Omega \subset \mathbb{R}^n$.*

Proof. Since Ω is bounded and (by translation) may be assumed to satisfy $0 \notin \Omega$, by Theorem 7.5, it suffices to establish the following lemma. □

Lemma 8.4 (Radial polynomial approximators). *Given $\rho > 0$ there exists $m \in \mathbb{N}$ such that*

$$\psi(x) := \frac{|x|^{m+1}}{m+1} \tag{8.6}$$

defines a strict approximator on $B_\rho(0) \setminus \{0\}$ for $\mathcal{M}_R$ and for $\mathcal{M}^R$.

Proof. Making use of the radial calculation of Remark 3.17(2), ψ has reduced 2-jet

$$(p, A) := (D\psi(x), D^2\psi(x)) = |x|^{m-1}(x, I + (m-1)P_x)$$
$$= |x|^{m-1}(x, P_{x^\perp} + mP_x), \tag{8.7}$$

where we recall that P_x is the orthogonal projection onto the line $[x]$ through $x \in \mathbb{R}^n \setminus \{0\}$ and $P_{x^\perp} = I - P_x$ is the orthogonal projection on $[x]^\perp$. Hence, the claim that ψ is a strict $\mathcal{M}$-approximator on $B_\rho(0) \setminus \{0\}$ is equivalent to the claim that

$$(x, I + (m-1)P_x) \in \text{Int } \mathcal{M} \quad \text{for every } x \text{ with } 0 < |x| < \rho. \tag{8.8}$$

Now we verify (8.8) for $\mathcal{M}_R$ and $\mathcal{M}^R$. Note that $I + (m-1)P_x = P_{x^\perp} + mP_x$ has $n-1$ eigenvalues equal to 1 and one eigenvalue equal to m.

For $\mathcal{M} = \mathcal{M}_R$, claim (8.8) becomes (since $I + (m-1)P_x > 0$)

$$m^{1/n} - \frac{|x|}{R} > 0,$$

which holds if and only if $m > (\rho/R)^n$.

For $\mathcal{M} = \mathcal{M}^R$, with arbitrary $e \in \mathbb{R}^n$ satisfying $|e| = 1$, claim (8.8) becomes

$$\langle (I + (m-1)P_x)e, e \rangle - \frac{|\langle x, e \rangle|}{R} > 0. \tag{8.9}$$

A simple calculation gives

$$\langle (I + (m-1)P_x)e, e \rangle = 1 + (m-1)\langle x/|x|, e \rangle^2,$$

and hence the needed (8.9) can be written as

$$g(t) := 1 - \frac{|x|t}{R} + (m-1)t^2 > 0 \quad \text{for } t := |\langle x/|x|, e \rangle| \geq 0. \tag{8.10}$$

The quadratic polynomial g takes on its minimum at $t_0 := \frac{|x|}{2R(m-1)}$, with minimum value

$$g(t_0) = 1 - \frac{|x|^2}{4R^2(m-1)} > 1 - \frac{\rho^2}{4R^2(m-1)} \quad \text{for all } x \text{ with } |x| < \rho.$$

Taking m sufficiently large gives $g(t_0) > 0$ and hence (8.10). $\qquad \square$

We remark that in the case of $\mathcal{M} = \mathcal{M}(R)$ one has

$$\lambda_{\min}(I + (m-1)P_x) = 1 > \frac{|x|}{R}$$

only if $\rho \leq R$, which leads to the restriction on domain size for this case.

Theorem 8.3 easily extends to nonreduced subequations (where the variable r is not silent) as follows. Recall that for $0 \leq \gamma < \infty$,

$$\mathcal{M}(\gamma) := \{(r, p, A) \in \mathcal{J}^2 : r \leq -\gamma|p|\}. \tag{8.11}$$

Theorem 8.5. *Suppose that $\mathcal{F} \subset \mathcal{J}^2$ is a subequation which is $\mathcal{M}$-monotone. If $\mathcal{M}$ contains either $\mathcal{M}(\gamma) \cap \mathcal{M}_R$ or $\mathcal{M}(\gamma) \cap \mathcal{M}^R$ for some γ, R, then comparison holds for $\mathcal{F}$ on all bounded domains $\Omega \subset\subset \mathbb{R}^n$.*

Proof. First, choose $m \in \mathbb{N}$ as in Lemma 8.4. Replace ψ in (8.6) by the radial polynomial

$$\psi(x) := \frac{|x|^{m+1}}{m+1} - C, \tag{8.12}$$

with $C > 0$ large to be determined. The reduced 2-jet remains unchanged but $r := \frac{|x|^{m+1}}{m+1} - C$. Since $|p| = |x|^m$, we have

$$r \leq -\gamma|p| \quad \Leftrightarrow \quad \frac{|x|^{m+1}}{m+1} + \gamma|x|^m \leq C.$$

This holds on the set where $|x| \leq \rho$ if $C \geq \frac{|\rho|^{m+1}}{m+1} + \gamma|\rho|^m$.   □

Remark 8.6. Theorem 8.3 also holds if $\mathcal{M}$ contains either of the following cones. For $\lambda, \Lambda \in \mathbb{R}$ with $0 < \lambda \leq \Lambda < +\infty$, define

$$\mathcal{M}^-_{\lambda,\Lambda,R} := \left\{ (r, p, A) \in \mathcal{J}^2 : \lambda \operatorname{tr} A^+ + \Lambda \operatorname{tr} A^- \geq \frac{\lambda n |p|}{R} \right\}, \qquad (8.13)$$

where $A = A^+ + A^-$ is the standard decomposition of A into its positive and negative parts. For $\delta \in \mathbb{R}$ with $\delta > 0$, define

$$\mathcal{M}(R)_\delta := \left\{ (r, p, A) \in \mathcal{J}^2 : \lambda_{\min}(A) + \delta \operatorname{tr} A \geq \frac{|p|}{R} \right\}. \qquad (8.14)$$

Proof. In both cases if suffices to check that the function ψ defined by (8.6) satisfies the claim (8.8) of Lemma 8.4. For $\mathcal{M}^-_{\lambda,\Lambda,R}$ the claim (8.8) becomes

$$\lambda(n + m - 1) - \frac{\lambda n |x|}{R} > 0. \qquad (8.15)$$

Since $|x| < \rho$, one has (8.15) if and only if $m > 1 + n\left(\frac{\rho}{R} - 1\right)$.

For $\mathcal{M}(R)_\delta$ the claim (8.8) becomes

$$1 + \delta(n - 1 + m) - \frac{|x|}{R} > 0. \qquad (8.16)$$

Since $|x| < \rho$, one has (8.16) if and only if $m > 1 - n + \frac{\rho - R}{\delta R}$.   □

Chapter Nine

<hr style="border: 2px solid black">

Failure of Comparison with Insufficient

Maximal Monotonicity

In this chapter we give some examples of subequation constraint sets $\mathcal{F}$ for which comparison fails to hold on a family of bounded domains Ω. Necessarily, this failure implies that the maximal monotonicity cone $\mathcal{F}_M$ for $\mathcal{F}$ does not contain any of the elements of our fundamental family with $R = +\infty$, nor does $\mathcal{F}_M$ contain any of the cones discussed in the previous chapter on additional monotonicity. We focus on two such examples. The first shows (as claimed in the introduction of Chapter 8) that Theorem 7.6 is sharp in the case when R is finite, that is, R gives an upper bound on the diameter of Ω for which comparison holds. The second shows just how bad the situation can be. Comparison can fail on arbitrarily small balls.

9.1 FINITE R AND FAILURE OF COMPARISON ON LARGE DOMAINS

We begin with a simple family of examples which illustrates the sharpness of Theorem 7.6 on the comparison principle in the case when R is finite.

Example 9.1. In dimension $n \geq 2$, with $k \in \{1, \ldots, n\}$ and $R \in (0, +\infty)$, consider the subequation constraint sets

$$\mathcal{F}_{k,R}^{\pm} := \{(r, p, A) \in \mathcal{J}^2 : \lambda_k(A) \pm \tfrac{|p|}{R} \geq 0\}, \tag{9.1}$$

where $\lambda_1(A) \leq \cdots \leq \lambda_n(A)$ are the ordered eigenvalues of $A \in \mathcal{S}(n)$. The constraints $\mathcal{F}_{k,R}^{\pm}$ define the subharmonics for the operators $F_{k,R}^{\pm}(Du, D^2u) = \lambda_k$ $(D^2u) \pm \tfrac{1}{R}|Du|$. One easily checks that each $\mathcal{F}_{k,R}^{\pm}$ is a subequation constraint set and that the following duality relations hold:

$$\widetilde{\mathcal{F}}_{k,R}^{+} = \mathcal{F}_{n+k-1,R}^{-} \quad \text{and} \quad \widetilde{\mathcal{F}}_{k,R}^{-} = \mathcal{F}_{n+k-1,R}^{+}. \tag{9.2}$$

Notice that two members of the family are subequation cones that we have seen before, namely

$$\mathcal{F}_{1,R}^{-} = \mathcal{M}(R) \quad \text{and} \quad \mathcal{F}_{n,R}^{+} = \widetilde{\mathcal{M}}(R), \tag{9.3}$$

where we note that only the first cone $\mathcal{M}(R)$ is convex.

Proposition 9.2. *For the family of subequations $\mathcal{F}_{k,R}^\pm$ in Example 9.1, one has the following statements:*

(a) *For each k and R, the maximal monotonicity cone $\mathcal{M}_{\mathcal{F}_{k,R}^\pm}$ of $\mathcal{F}_{k,R}^\pm$ equals*

$$\mathcal{M}(R) := \{(s,q,B) \in \mathcal{J}^2 : B \geq \tfrac{|q|}{R} I\}$$
$$= \{(s,q,B) \in \mathcal{J}^2 : \lambda_1(B) - \tfrac{|q|}{R} \geq 0\}.$$

Consequently, comparison holds for each $\mathcal{F}_{k,R}^\pm$ on $\overline{\Omega}$ for every domain Ω contained in a ball B_R of radius R.

(b) *For each $k = 2, \ldots, n$, comparison fails for $\mathcal{F}_{k,R}^\pm$ on any ball $B_{R'}$ with radius $R' > R$.*

Proof. It suffices to consider the family of subequations $\mathcal{F}_{k,R}^+$ since each $\mathcal{F}_{k,R}^-$ is dual to $\mathcal{F}_{n-k+1,R}^+$. This is because dual subequations have the same maximal monotonicity cone (see Proposition 4.5(b)) and the comparison principle holds for a subequation $\mathcal{F}$ if and only if it holds for its dual subequation $\widetilde{\mathcal{F}}$.

We begin by showing that each $\mathcal{F}_{k,R}^+$ is $\mathcal{M}(R)$-monotone. Given any $(r,p,A) \in \mathcal{F}_{k,R}^+$ and any $(s,q,B) \in \mathcal{M}(R)$, making use of the *dual Weyl inequality*

$$\lambda_k(A+B) \geq \lambda_k(A) + \lambda_1(B) \quad \text{for each } A,B \in \mathcal{S}(n), \ k = 1, \ldots, n,$$

the triangle inequality on $\mathbb{R}^n$, and using $\lambda_1(B) \geq |q|/R$, one has

$$\lambda_k(A+B) + \frac{|p+q|}{R} \geq \lambda_k(A) + \lambda_1(B) + \frac{|p|}{R} - \frac{|q|}{R} \geq 0.$$

Hence, we have $\mathcal{M}(R) \subset \mathcal{M}_{\mathcal{F}_{k,R}^+}$, the maximal monotonicity cone (as defined in Definition 4.2). It remains to check the reverse inclusion, that is,

$$(r+s, p+q, A+B) \in \mathcal{F}_{k,R}^+ \ \forall (r,p,A) \in \mathcal{F}_{k,R}^+ \ \Rightarrow \ (s,q,B) \in \mathcal{M}(R). \quad (9.4)$$

Since both $\mathcal{F}_{k,R}^+$ and $\mathcal{M}(R)$ are reduced subequations, condition (9.4) can be written as

$$\lambda_k\left(A+B+\frac{|p+q|}{R}I\right) \geq 0 \ \forall (p,A): \ \lambda_k\left(A+\frac{|p|}{R}I\right) \geq 0 \ \Rightarrow \ B \geq \frac{|q|}{R}I. \quad (9.5)$$

We will use the fact that

$$\lambda_k(A+P) \geq 0 \ \forall A \in \mathcal{S}(n): \ \lambda_k(A) \geq 0 \ \Rightarrow \ P \geq 0; \quad (9.6)$$

that is, the maximal monotonicity cone for $\{A \in \mathcal{S}(n) : \lambda_k(A) \geq 0\}$ is $\mathcal{P}$. Let $(q, B) \in \mathbb{R}^n \times \mathcal{S}(n)$ satisfy the hypothesis in (9.5), which is equivalent to

$$\lambda_k\left(D + B + \frac{|p+q| - |p|}{R}I\right) \geq 0 \;\; \forall (p, D): \;\; \lambda_k(D) \geq 0 \;\; \Rightarrow \;\; B \geq \frac{|q|}{R}I. \quad (9.7)$$

By applying (9.6) to the pair (q, B) satisfying (9.7), one finds

$$B + \frac{|p+q| - |p|}{R}I \geq 0 \quad \text{for every } p \in \mathbb{R}^n,$$

which yields $B - \frac{|q|}{R}I \geq 0$ with $p = -2q$. This completes the proof that $\mathcal{M}(R)$ is the maximal monotonicity cone for each $\mathcal{F}^+_{k,R}$.

It then follows from Theorem 7.6 that comparison holds for all domains Ω contained in a ball B_R of radius R.

Next we note that the same radial counterexample $u(x) := |x| - \frac{|x|^2}{2R}$ to the (MP) for $\widetilde{\mathcal{M}}(R) = \mathcal{F}^+_{n,R}$ on balls $B_{R'}$ with radius $R' > R$ (see Proposition 6.5) is a counterexample to the (MP) for the reduced subequation $\mathcal{F}^+_{k,R}$ on $B_{R'}$ if $k \geq 2$. This is because $D^2 u(x)$ has $n-1$ eigenvalues equal to $-\frac{1}{R} + \frac{1}{|x|}$, which are all greater than the remaining eigenvalue $-\frac{1}{R}$ (see formula (6.17)). Hence, with $(p, A) := (Du(x), D^2 u(x))$, for each $x \neq 0$,

$$\lambda_k(A) + \frac{|p|}{R} = \lambda_n(A) + \frac{|p|}{R} \quad \text{if } k \geq 2, \quad (9.8)$$

which shows that the $\widetilde{\mathcal{M}}(R) = \mathcal{F}^+_{n,R}$-subharmonic function u is also $\mathcal{F}^+_{k,R}$-subharmonic on $\mathbb{R}^n \setminus \{0\}$ for $k \geq 2$. Recall that u is trivially $\widetilde{\mathcal{M}}(R)$-subharmonic at the origin because there are no upper test jets at the origin, and hence the same claim for $\mathcal{F}^+_{k,R}$.

Finally, since $\mathcal{F}^+_{k,R}$ is a reduced subequation cone, the constant function defined by $w \equiv \sup_\Omega u$ is $\mathcal{F}^+_{k,R}$-harmonic (superharmonic) and hence the failure of the (MP) implies the failure of comparison. $\quad\square$

We remark that Example 9.1 is a special case of a larger family of counterexamples to the validity of comparison principles and Alexandroff estimates. See [39, Section 4] for operators involving truncated Laplacians and truncated Pucci maximal and minimal operators. Moreover, Proposition 9.2 easily generalizes with the standard eigenvalues $\lambda_k(A)$ replaced by the Gårding eigenvalues $\lambda^{\mathfrak{g}}_k(A)$ of a *Gårding–Dirichlet polynomial* $\mathfrak{g}$ (see the discussion of Section 11.6 for the relevant notions).

Example 9.3. Let $\mathfrak{g}$ be a Gårding–Dirichlet polynomial of degree m on $\mathcal{S}(n)$ in the sense of Definition 11.31 whose (*ordered*) *Gårding I-eigenvalues of A* are denoted by

$$\lambda_1^{\mathfrak{g}}(A) \le \lambda_2^{\mathfrak{g}}(A) \le \cdots \le \lambda_m^{\mathfrak{g}}(A), \tag{9.9}$$

and $\mathfrak{g}$ is normalized to have $\mathfrak{g}(I) = 1$. Again assuming that $n \ge 2$, the same conclusions as Proposition 9.2 hold for the subequation constraint set

$$\mathcal{F}_{k,\beta}^{\mathfrak{g}} := \{(r,p,A) \in \mathcal{J}^2 : \lambda_k^{\mathfrak{g}}(A) + \beta|p| \ge 0\}, \quad \text{where } k \in \{2,\dots,n\}. \tag{9.10}$$

9.2   FAILURE OF COMPARISON ON ARBITRARILY SMALL DOMAINS

We now give a family of examples for which comparison fails on arbitrarily small balls. In fact, we will exhibit subequations for which existence of the Dirichlet problem will hold on arbitrary balls, but the comparison principle, the maximum principle, and uniqueness for the Dirichlet problem will all fail (see Theorem 9.8 below). The argument will make use of the considerations of Appendix A on maximal and minimal solutions, and hence the proof of Theorem 9.8 will be given in Appendix B. The examples we present will involve subequations $\mathcal{F}$ whose maximal monotonicity cone $\mathcal{M}_{\mathcal{F}}$ has empty interior, and, as such, cannot admit strict approximators on any domain, no matter how small. The examples are *reduced* in the sense that no constraint is placed on the jet variable $r \in \mathbb{R}$ and hence the subequation $\mathcal{F}$ will be considered as subsets of $\mathbb{R}^n \times \mathcal{S}$ (see Chapter 10 for more on reductions).

We begin by defining the subequations and making some preliminary observations. For $p \ne 0$ in $\mathbb{R}^n$, we recall that the orthogonal projections onto the subspaces $[p]$ and $[p]^{\perp}$ are (respectively)

$$P_p = \frac{1}{|p|^2} p \otimes p \quad \text{and} \quad P_{p^{\perp}} = I - P_p. \tag{9.11}$$

Example 9.4. For $\alpha \in (1, +\infty)$, define

$$B(p,A) := A + |p|^{\frac{\alpha-1}{\alpha}}(P_{p^{\perp}} + \alpha P_p)) \text{ if } p \ne 0 \quad \text{and} \quad B(0,A) := A. \tag{9.12}$$

Notice that the map $B \colon \mathbb{R}^n \times \mathcal{S}(n) \to \mathcal{S}(n)$ is continuous. Consider the operators $F, G \in C(\mathbb{R}^n \times \mathcal{S}(n))$ defined by

$$F(p,A) := \lambda_{\min}(B(p,A)) \quad \text{and} \quad G(p,A) := \lambda_{\max}(B(p,A)), \tag{9.13}$$

along with the (reduced) subequations defined by

$$\mathcal{F} := \{(p,A) \in \mathbb{R}^n \times \mathcal{S}(n) : \lambda_{\min}(B(p,A)) \ge 0\} \tag{9.14}$$

and

$$\mathcal{G} := \left\{ (p, A) \in \mathbb{R}^n \times \mathcal{S}(n) : \lambda_{\max}(B(p, A)) \geq 0 \right\}. \tag{9.15}$$

When $\alpha = 2$, the subequation $\mathcal{F}$ was introduced in [49] as an example where existence holds, but uniqueness fails. Hence, we are considering generalizations of that example.

The fact that the closed sets $\mathcal{F}$ and $\mathcal{G}$ are (reduced) subequations can be seen as follows. Property (N) is automatic since $\mathcal{F}$ and $\mathcal{G}$ are independent of the jet variable $r \in \mathbb{R}$. Property (P) holds for $\mathcal{F}$ and $\mathcal{G}$ since the operators F and G are increasing in $A \in \mathcal{S}(n)$.

To prove the topological property (T) and to show compatibility between F and $\mathcal{F}$ and between G and $\mathcal{G}$, we use a general lemma which we state in the reduced case. By *compatibility* we mean the relation (9.17) below (see Definition 11.1). This notion will play an important role in our treatment of comparison for operators F.

Lemma 9.5. *Suppose that $F \in C(\mathbb{R}^n \times \mathcal{S}(n))$ is a degenerate elliptic operator, that is, $F = F(p, A)$ is increasing in A on all of $\mathcal{S}(n)$. If F is linear on lines through $I \in \mathcal{S}(n)$ in the sense that*

$$F(p, A + tI) = F(p, A) + t \quad \text{for each } t \in \mathbb{R},$$

then the set $\mathcal{F} := \{ (p, A) \in \mathbb{R}^n \times \mathcal{S}(n) : F(p, A) \geq 0 \}$ satisfies

$$\operatorname{Int} \mathcal{F} = \left\{ (p, A) \in \mathbb{R}^n \times \mathcal{S}(n) : F(p, A) > 0 \right\}, \tag{9.16}$$

$$\partial \mathcal{F} = \left\{ (p, A) \in \mathbb{R}^n \times \mathcal{S}(n) : F(p, A) = 0 \right\}, \tag{9.17}$$

$$\sim \mathcal{F} = \left\{ (p, A) \in \mathbb{R}^n \times \mathcal{S}(n) : F(p, A) < 0 \right\}. \tag{9.18}$$

Proof. Since $\{ (p, A) : F(p, A) > 0 \}$ is an open subset of $\mathcal{F}$, it is contained in $\operatorname{Int} \mathcal{F}$. If $F(p, A) = 0$, then (p, A) is approximated by $(p, A + \varepsilon I) \in \operatorname{Int} \mathcal{F}$ for each $\varepsilon > 0$ since $F(p, A + \varepsilon I) = F(p, A) + \varepsilon = \varepsilon > 0$. Such a (p, A) is also approximated by $(p, A - \varepsilon I) \notin \mathcal{F}$ for each $\varepsilon > 0$ since $F(p, A - \varepsilon I) = F(p, A) - \varepsilon = -\varepsilon < 0$. This proves (9.16), (9.17), and (9.18), as well as property (T): $\mathcal{F} = \overline{\operatorname{Int} \mathcal{F}}$. $\qquad \square$

This result applies to both F and G defined by (9.13) above since $B(p, A + tI) = B(p, A) + tI$ and $\lambda_j(B(p, A) + tI) = \lambda_j(B(p, A)) + t$ for each $j = 1, \ldots, n$. The Dirichlet duals of $\mathcal{F}$ and $\mathcal{G}$ defined in (9.14) and (9.15) are given by

$$\widetilde{\mathcal{F}} := \left\{ (p, A) \in \mathbb{R}^n \times \mathcal{S}(n) : \lambda_{\max} A - |p|^{\frac{\alpha-1}{\alpha}} (P_{p^\perp} + \alpha P_p)) \geq 0 \right\} \tag{9.19}$$

and

$$\widetilde{\mathcal{G}} := \left\{ (p, A) \in \mathbb{R}^n \times \mathcal{S}(n) : \lambda_{\min}(A - |p|^{\frac{\alpha-1}{\alpha}} (P_{p^\perp} + \alpha P_p)) \geq 0 \right\}. \tag{9.20}$$

Note that one has the inclusions

$$\mathcal{F} \subset \mathcal{G} \quad \text{and} \quad \tilde{\mathcal{G}} \subset \tilde{\mathcal{F}} \tag{9.21}$$

and that

$$(p, A) := (0, 0) \in \mathcal{F} \cap \tilde{\mathcal{F}} \cap \mathcal{G} \cap \tilde{\mathcal{G}}. \tag{9.22}$$

Remark 9.6. If $\alpha = 1$ is considered, then

$$\mathcal{F} = \mathbb{R}^n \times (\mathcal{P} - I), \quad \tilde{\mathcal{F}} = \mathbb{R}^n \times (\tilde{\mathcal{P}} + I), \quad \mathcal{G} = \mathbb{R}^n \times (\tilde{\mathcal{P}} - I), \quad \tilde{\mathcal{G}} = \mathbb{R}^n \times (\mathcal{P} + I)$$

are all pure second order, so comparison holds for all of them and Int $\mathcal{P}$ governs boundary convexity, as it is the asymptotic interior of each of them (see the discussion in Appendix A).

Next we describe the maximal monotonicity cones for the two subequations in this Example 9.4.

Lemma 9.7. *For each $\alpha \in (1, +\infty)$ the (reduced) subequations*

$$\mathcal{F} := \left\{ (p, A) \in \mathbb{R}^n \times \mathcal{S}(n) : \lambda_{\min}(B(p, A)) \geq 0 \right\}$$

and

$$\mathcal{G} := \left\{ (p, A) \in \mathbb{R}^n \times \mathcal{S}(n) : \lambda_{\max}(B(p, A)) \geq 0 \right\},$$

where

$$B(p, A) := A + |p|^{\frac{\alpha-1}{\alpha}} (P_{p^\perp} + \alpha P_p)) \ \textit{if} \ p \neq 0 \quad \textit{and} \quad B(0, A) := A,$$

have (reduced) maximal monotonicity cone $\mathcal{M} = \{0\} \times \mathcal{P} \subset \mathbb{R}^n \times \mathcal{S}(n)$.

Proof. By Proposition 4.5(b), $\mathcal{F}$ and $\tilde{\mathcal{F}}$ have the same maximal monotonicity cones. We will show that $\mathcal{M}_{\tilde{\mathcal{F}}} = \{0\} \times \mathcal{P}$. It suffices to show that if $(q, B) \in \mathcal{M}_{\tilde{\mathcal{F}}}$ then q must be equal to 0 because the fiber of $\tilde{\mathcal{F}}$ over $\{0\}$ is $\tilde{\mathcal{P}}$, which has maximal monotonicity $\mathcal{P}$. Suppose that $(q, B) \in \mathcal{M}_{\tilde{\mathcal{F}}}$; then since $\mathcal{M}_{\tilde{\mathcal{F}}}$ satisfies property (P), one has

$$(q, \lambda I) \in \mathcal{M}_{\tilde{\mathcal{F}}} \quad \text{if} \ \lambda I \geq B.$$

Since $\mathcal{M}_{\tilde{\mathcal{F}}}$ is a monotonicity cone for $\tilde{\mathcal{F}}$ with $(0, 0) \in \tilde{\mathcal{F}}$, one has for any $\lambda \in \mathbb{R}$ such that $\lambda I \geq B$,

$$(tq, t\lambda I) = (0, 0) + (tq, t\lambda I) \in \tilde{\mathcal{F}} + \mathcal{M}_{\tilde{\mathcal{F}}} \subset \tilde{\mathcal{F}} \quad \text{for all} \ t > 0,$$

that is,

$$\lambda_{\max}\left(t\lambda I - t^{\frac{\alpha-1}{\alpha}}|q|^{\frac{\alpha-1}{\alpha}}(P_{q^\perp} + \alpha P_q)\right) \geq 0 \quad \text{for all } t > 0. \tag{9.23}$$

Pick $\lambda > 0$ large enough to ensure that $\lambda I \geq B$. Since the eigenvalues of $P_{q^\perp} + \alpha P_q$ are 1 and $\alpha > 1$, and since $(\alpha - 1)/\alpha = 1 - 1/\alpha$, inequality (9.23) is equivalent to

$$\lambda \geq \frac{|q|^{\frac{\alpha-1}{\alpha}}}{t^{1/\alpha}} \quad \text{for all } t > 0,$$

which implies $q = 0$, as desired.

An analogous argument shows that $\mathcal{M}_{\mathcal{G}} = \mathcal{M}_{\widetilde{\mathcal{G}}} = \{0\} \times \mathcal{P}$. $\qquad\square$

Notice that the interiors of the maximal monotonicity cones are empty, and hence strict approximators cannot be found, which suggests that comparison may fail. Indeed, comparison does fail.

Theorem 9.8. *Let $R \in (0, +\infty)$ and let $B_R \subset \mathbb{R}^n$ be the open R-ball about 0. The functions $z \in C^\infty(\mathbb{R}^n)$ and $h \in C^{2,\alpha-1}(\mathbb{R}^n)$ defined by*

$$z(x) := 0 \quad \text{and} \quad h(x) := -\frac{|x|^{1+\alpha}}{1+\alpha} + \frac{R^{1+\alpha}}{1+\alpha}, \quad x \in \mathbb{R}^n$$

are both $\mathcal{F}$- and $\mathcal{G}$-harmonic on all of $\mathbb{R}^n$. They both have boundary values $\varphi = 0$ on ∂B_R. Thus comparison, uniqueness, and the maximum principle all fail for $\mathcal{F}$ and $\mathcal{G}$ on B_R, which can be an arbitrarily small ball.

The direct proof is provided, for the convenience of the reader, in Appendix B, where we compute the one-variable radial subequations associated to $\mathcal{F}$ and to $\mathcal{G}$.

Chapter Ten

Special Cases: Reduced Constraint Sets

In Theorem 7.5 we gave a general comparison principle on domains $\Omega \subset\subset \mathbb{R}^n$ for subequation constraint sets $\mathcal{F} \subset \mathcal{J}^2 = \mathbb{R} \times \mathbb{R}^n \times \mathcal{S}(n)$ which satisfy the assumptions

$$\mathcal{F} \text{ is } \mathcal{M}\text{-monotone with } \mathcal{M} \text{ a monotonicity cone subequation,} \qquad (10.1)$$

$$\mathcal{M} \text{ admits a strict approximator on } \Omega. \qquad (10.2)$$

Moreover, in Definition 5.2 we introduced the family of $(\gamma, \mathcal{D}, R)$-monotonicity cone subequations and we described for which domains Ω a given cone $\mathcal{M}(\gamma, \mathcal{D}, R)$ admits a strict approximator ψ on Ω.

In this chapter we discuss the special cases when at least one of the constraint factors of $\mathcal{F} \subset \mathbb{R} \times \mathbb{R}^n \times \mathcal{S}(n)$ is "silent" in the sense that no restriction is placed on the jet variable corresponding to that factor.[1] In these cases, the subset of the remaining factors will be called the *reduced constraint set for $\mathcal{F}$* and will be denoted by $\mathcal{F}'$. The silent factors can be included in any associated monotonicity cone subequation $\mathcal{M}$ for $\mathcal{F}$, making $\mathcal{M}$ "large" and hence increasing the likelihood of (10.2) being true.

We start with the *pure second-order case*. Although it has been treated in some detail in [46], a discussion is in order here as a prelude to the main results of this chapter which concern the *gradient-free* case, where we prove the analogue of the *subaffine theorem*. We apply Convention 3.11 throughout this chapter.

At this point, some readers may wish to turn to the Summary Remark 10.14 for an overview.

10.1 PURE SECOND ORDER

By definition, a *pure second-order constraint set* is a subset $\mathcal{F} \subset \mathcal{J}^2 = \mathbb{R} \times \mathbb{R}^n \times \mathcal{S}(n)$ of the form

$$\mathcal{F} = \mathbb{R} \times \mathbb{R}^n \times \mathcal{F}'.$$

---

[1] In terms of nonlinear differential operators, this means that $F(u, Du, D^2 u)$ is independent of at least one of the variables u, Du, or $D^2 u$.

That is, the factor $\mathbb{R} \times \mathbb{R}^n$ is silent and the reduced constraint set $\mathcal{F}'$ is a subset of $\mathcal{S}(n)$. An equivalent definition in terms of monotonicity is that $\mathcal{F}$ is $\mathbb{R} \times \mathbb{R}^n \times \{0\}$-monotone, that is,

$$(r, p, A) \in \mathcal{F} \;\Rightarrow\; (r + s, p + q, A) \in \mathcal{F} \text{ for each } s \in \mathbb{R}, \ q \in \mathbb{R}^n. \tag{10.3}$$

In particular, $\mathcal{F}$ is $(-\infty, 0] \times \mathbb{R}^n \times \{0\}$-monotone, so that $\mathcal{F}$ automatically satisfies both the negativity property (N) and the directionality property (D) with respect to the cone $\mathcal{D}$, which is all of $\mathbb{R}^n$. The positivity property (P) holds for $\mathcal{F}$ if and only if the reduced constraint set $\mathcal{F}'$ is $\mathcal{P}$-monotone, that is,

$$\mathcal{F}' + \mathcal{P} \subset \mathcal{F}'. \tag{10.4}$$

The remaining subequation property, namely the topological property (T) which asks that $\mathcal{F}$ is the closure of its interior, is equivalent to this being true for the reduced constraint set $\mathcal{F}'$, that is,

$$\mathcal{F}' = \overline{\text{Int } \mathcal{F}'}. \tag{10.5}$$

This topological property (10.5) follows from the positivity property (10.4) as long as $\mathcal{F}$ (or equivalently $\mathcal{F}'$) is closed. To see this, first note that the positivity property (10.4) implies that $\mathcal{F}' + \text{Int } \mathcal{P} \subset \text{Int } \mathcal{F}'$. In particular, if $A \in \mathcal{F}'$, then $A + \varepsilon I \in \text{Int } \mathcal{F}'$ for every $\varepsilon > 0$. Consequently, $\mathcal{F} \subset \mathcal{J}^2$ is a *pure second-order subequation* if

$$\mathcal{F} \text{ is closed and } \mathcal{M}(\mathcal{P}) = \mathbb{R} \times \mathbb{R}^n \times \mathcal{P}\text{-monotone.} \tag{10.6}$$

In terms of the reduced constraint set $\mathcal{F}'$ with $\mathcal{F} = \mathbb{R} \times \mathbb{R}^n \times \mathcal{F}'$ the definition takes the simpler equivalent form

$$\mathcal{F}' \subset \mathcal{S}(n) \text{ is closed and } \mathcal{P}\text{-monotone.} \tag{10.7}$$

Employing Convention 3.11, we will refer to such an $\mathcal{F}'$ as a *pure second-order subequation*.[2] Taking $\mathcal{F}' \subsetneq \mathcal{S}(n)$ as a proper subset ensures that $\mathcal{F}(X)$ is a proper subset of $\text{USC}(X)$.

Comparison for arbitrary bounded domains Ω follows easily from Theorem 7.6 since the monotonicity cone $\mathcal{M}(\mathcal{P}) = \mathbb{R} \times \mathbb{R}^n \times \mathcal{P}$ for pure second-order subequations contains the monotonicity cone $\mathcal{M}(\mathcal{N}) \cap \mathcal{M}(\mathcal{P}) = \mathcal{N} \times \mathbb{R}^n \times \mathcal{P}$. Note that this is a special case of a $(\gamma, \mathcal{D}, R)$-monotonicity cone with $\gamma = 0$, $\mathcal{D} = \mathbb{R}^n$, and $R = +\infty$.

---

[2]In [77], open sets $\Theta \subsetneq \mathcal{S}(n)$ which correspond to Int $\mathcal{F}'$ are called *elliptic sets*. Pure second-order subequations $\mathcal{F}'$ are introduced as closed subsets of $\mathcal{S}(n)$ with the positivity property in [46], where they are denoted by F and called *Dirichlet sets*. Such $\mathcal{F}'$ are called *elliptic sets* and denoted by Θ in [23].

The differential inclusion (2.7) defining $\mathcal{F}$-subharmonicity at x_0 is

$$J^2_{x_0}\varphi = (\varphi(x_0), D\varphi(x_0), D^2\varphi(x_0)) \in \mathcal{F} \qquad (10.8)$$

for each C^2-upper test function φ for u at x_0. This reduces to the simpler statement that $u \in \mathcal{F}(X)$ if and only if $u \in \mathrm{USC}(X)$ and, for each $x_0 \in X$, one has

$$D^2\varphi(x_0) \in \mathcal{F}' \quad \text{for each } C^2\text{-upper test function } \varphi \text{ for } u \text{ at } x_0. \qquad (10.9)$$

Example 10.1. What might be considered the most basic example in all of viscosity theory is the example $\mathcal{F}' = \mathcal{P}$. Its dual is the *subaffine subequation*

$$\widetilde{\mathcal{P}} := \{A \in \mathcal{S}(n) : \lambda_{\max}(A) \geq 0\}. \qquad (10.10)$$

One has the following elementary facts:

$$\mathcal{F} \text{ is a pure second-order subequation if and only if } \widetilde{\mathcal{F}} \text{ is} \qquad (10.11)$$

and

$$\mathcal{F}' + \widetilde{\mathcal{F}}' \subset \widetilde{\mathcal{P}} \text{ for each pure second-order subequation } \mathcal{F} = \mathbb{R} \times \mathbb{R}^n \times \mathcal{F}'. \qquad (10.12)$$

This pure second-order jet addition theorem (10.12) implies the subharmonic addition theorem

$$\mathcal{F}'(X) + \widetilde{\mathcal{F}}'(X) \subset \widetilde{\mathcal{P}}(X), \qquad (10.13)$$

as discussed in Theorem 7.1 for general sums $\mathcal{F} + \mathcal{G} \subset \mathcal{H}$ of subequations in $\mathcal{J}^2$. Let Aff denote the space of affine functions on $\mathbb{R}^n$. As defined in Example 3.9, for each open set $X \subset \mathbb{R}^n$, let $\mathrm{SA}(X)$ denote the set of all $u \in \mathrm{USC}(X)$ with the *subaffine property*

$$u \leq a \text{ on } \partial\Omega \implies u \leq a \text{ on } \Omega \quad \text{for every } \Omega \subset\subset X \text{ and } a \in \mathrm{Aff}. \qquad (10.14)$$

One of the key results of [46] is that

$$\widetilde{\mathcal{P}}(X) = \mathrm{SA}(X). \qquad (10.15)$$

The proof of (10.15) is also obtained here as a special, easier case of the proof of Theorem 10.7, as indicated in its proof. Combining (10.15) with (10.13), one has the *subaffine theorem* of [46]:

$$\mathcal{F}(X) + \widetilde{\mathcal{F}}(X) \subset \mathrm{SA}(X). \qquad (10.16)$$

This gives rise to an alternate proof of the comparison principle for pure second-order subequations, because subaffine functions clearly satisfy the (ZMP) as the function zero is affine.

10.2   GRADIENT-FREE

By definition, a *gradient-free constraint set* is a subset $\mathcal{F} \subset \mathcal{J}^2 = \mathbb{R} \times \mathbb{R}^n \times \mathcal{S}(n)$ of the form

$$\mathcal{F} = \left\{(r, p, A) \in \mathcal{J}^2 : p \in \mathbb{R}^n \text{ and } (r, A) \in \mathcal{F}'\right\}, \quad \text{where } \mathcal{F}' \subset \mathbb{R} \times \mathcal{S}(n). \quad (10.17)$$

That is, the factor $\mathbb{R}^n$ is silent and the reduced constraint set $\mathcal{F}'$ is a subset of $\mathbb{R} \times \mathcal{S}(n)$. When it is convenient, we will reorder the jet variables as (p, r, A) so that the definition of $\mathcal{F}$ being gradient-free can be restated as

$$\mathcal{F} = \mathbb{R}^n \times \mathcal{F}' \quad \text{with } \mathcal{F}' \subset \mathbb{R} \times \mathcal{S}(n). \quad (10.18)$$

An equivalent definition in terms of monotonicity is that $\mathcal{F}$ is $\{0\} \times \mathbb{R}^n \times \{0\}$-monotone, that is,

$$(r, p, A) \in \mathcal{F} \implies (r, p+q, A) \in \mathcal{F} \text{ for each } q \in \mathbb{R}^n. \quad (10.19)$$

This is the directionality property (D) with the convex cone $\mathcal{D} \subset \mathbb{R}^n$ taken to be all of $\mathbb{R}^n$. Incorporating the subequation properties, we arrive at the definition of the concept we seek.

Definition 10.2. A *gradient-free subequation* is any closed, nonempty subset $\mathcal{F} \subsetneq \mathcal{J}^2 = \mathbb{R} \times \mathbb{R}^n \times \mathcal{S}(n)$, which is $\mathcal{M}$-monotone with respect to the monotonicity cone subequation $\mathcal{M}(\mathcal{N}) \cap \mathcal{M}(\mathcal{P}) = \mathcal{N} \times \mathbb{R}^n \times \mathcal{P}$, that is,

$$\mathcal{F} + (\mathcal{N} \times \mathbb{R}^n \times \mathcal{P}) \subset \mathcal{F}. \quad (10.20)$$

It is obvious that this monotonicity property is equivalent to the combined monotonicity properties (P), (N), and (D) with $\mathcal{D} = \mathbb{R}^n$. Therefore, in the gradient-free case, one always has comparison on arbitrary bounded domains by applying Theorem 7.6.

Theorem 10.3. *Let $\mathcal{F} \subset \mathcal{J}^2$ be a gradient-free subequation and let $\Omega \subset\subset \mathbb{R}^n$ be arbitrary. Given $u, v \in \mathrm{USC}(\overline{\Omega})$ which are $\mathcal{F}$-, $\widetilde{\mathcal{F}}$-subharmonic on Ω, one has*

$$u + v \leq 0 \text{ on } \partial\Omega \implies u + v \leq 0 \text{ on } \Omega.$$

In analogy with the pure second-order case, all of the above discussion can be reformulated in simpler terms by starting with the reduced constraint set $\mathcal{F}' \subset \mathbb{R} \times \mathcal{S}(n)$ and proceeding as follows.

Definition 10.4. A closed subset $\mathcal{F}' \subset \mathbb{R} \times \mathcal{S}(n)$ is called a *gradient-free subequation* if

$$\mathcal{F}' + \mathcal{Q} \subset \mathcal{F}', \quad \text{where } \mathcal{Q} := \mathcal{N} \times \mathcal{P}. \tag{10.21}$$

We note that the topological property (T) is again automatic since each $(r, A) \in \mathcal{F}'$ can be approximated by $(r, A) + \varepsilon(-1, I)$ which lies in $\mathcal{F}' + \operatorname{Int} \mathcal{Q} \subset \operatorname{Int} \mathcal{F}'$.

Examples 10.5. The most basic example is $\mathcal{F}' = \mathcal{Q} = \mathcal{N} \times \mathcal{P}$. Another example is its dual

$$\widetilde{\mathcal{Q}} := \big\{ (r, A) \in \mathbb{R} \times \mathcal{S}(n) : r \leq 0 \text{ or } A \in \widetilde{\mathcal{P}} \big\} \tag{10.22}$$

with $\widetilde{\mathcal{P}}$ as in (10.10), the subaffine subequation.

One has the following elementary facts:

$$\mathcal{F} \text{ is a gradient-free subequation if and only if } \widetilde{\mathcal{F}} \text{ is} \tag{10.23}$$

and

$$\mathcal{F}' + \widetilde{\mathcal{F}}' \subset \widetilde{\mathcal{Q}} \text{ for each gradient-free subequation } \mathcal{F} \text{ determined by } \mathcal{F}'. \tag{10.24}$$

This jet addition result extends to the *subharmonic addition theorem*

$$\mathcal{F}(X) + \widetilde{\mathcal{F}}(X) \subset \widetilde{\mathcal{Q}}(X) \quad \text{for any open subset } X \subset \mathbb{R}^n. \tag{10.25}$$

Here, in the spirit of Convention 3.11, one has $u \in \mathcal{F}(X)$ if and only if $u \in \mathrm{USC}(X)$ and, for each $x_0 \in X$, one has

$$(\varphi(x_0), D^2\varphi(x_0)) \in \mathcal{F}' \text{ for each } C^2\text{-upper test function } \varphi \text{ for } u \text{ at } x_0.$$

The claims up to (10.24) are purely algebraic statements and easily verified. Claim (10.25) follows (10.24) and from the general results of Chapter 7. In particular, see Theorem 7.1, Lemma 7.3, and Theorem 7.4.

We will finish this discussion by providing alternate characterizations of $\widetilde{\mathcal{Q}}(X)$, including the one claimed in Example 3.12, culminating in the *subaffine-plus theorem* (see Theorem 10.8 below). The key notion is that of *subaffine-plus functions*, which we proceed to describe.

For any domain $\Omega \subset\subset \mathbb{R}^n$ and with Aff the space of affine functions on $\mathbb{R}^n$, denote by

$$\text{Aff}^+(\overline{\Omega}) := \{\alpha := a|_{\overline{\Omega}} : a \in \text{Aff and } a \geq 0 \text{ on } \overline{\Omega}\}, \qquad (10.26)$$

the space of *affine-plus functions on* $\overline{\Omega}$.

Definition 10.6. For $X \subseteq \mathbb{R}^n$ open, a function $u \in \text{USC}(X)$ is said to be *subaffine plus on* X if for every open subset $\Omega \subset\subset X$, it has the *subaffine-plus property*:

$$u \leq \alpha \text{ on } \partial\Omega \;\Rightarrow\; u \leq \alpha \text{ on } \Omega \quad \text{for every } \alpha \in \text{Aff}^+(\overline{\Omega}). \qquad (10.27)$$

Denote by $\text{SA}^+(X)$ the space of all such functions.

The $\widetilde{\mathcal{Q}}$-subharmonic functions u on X can be characterized by being subaffine plus on X, and/or by the positive part $u^+ := \max\{u, 0\}$ being subaffine on X.

Theorem 10.7 (Subaffine-plus characterizations). *If $X \subseteq \mathbb{R}^n$ is open, then*

$$\widetilde{\mathcal{Q}}(X) = \text{SA}^+(X) = \{u \in \text{USC}(X) : u^+ \in \text{SA}(X) = \widetilde{\mathcal{P}}(X)\}. \qquad (10.28)$$

Proof. The proof requires verifying three inclusions, where the first inclusion provides motivation for the concept of *subaffine plus* introduced in Definition 10.6. For completeness, we include a proof of the characterization $\widetilde{\mathcal{P}}(X) = \text{SA}(X)$, which was stated in (10.15) and used in the statement of Theorem 10.7.

Step 1: $\widetilde{\mathcal{Q}}(X) \subset \text{SA}^+(X)$. Suppose that $u \in \widetilde{\mathcal{Q}}(X)$. Since $\widetilde{\mathcal{Q}}$ is a closed set satisfying (P) and (N) one has *definitional comparison* (see Lemma 3.14), that is,

$$u \leq w \text{ on } \partial\Omega \;\Rightarrow\; u \leq w \text{ on } \Omega \qquad (10.29)$$

for functions w that satisfy

$$w \in C(\overline{\Omega}) \cap C^2(\Omega) \quad \text{with } (w(x), D^2 w(x)) \notin \widetilde{\mathcal{Q}} \text{ for all } x \in \Omega. \qquad (10.30)$$

If w is a degree-2 polynomial then $D_x^2 w(x) := A$ is independent of x, so the requirement "$(w(x), D^2 w(x)) \notin \widetilde{\mathcal{Q}}$ for all $x \in \Omega$" in (10.30) becomes

$$w(x) > 0 \text{ and } D^2 w(x) := A < 0 \quad \text{for each } x \in \Omega. \qquad (10.31)$$

Such functions w can be constructed by starting from any $\alpha \in \text{Aff}^+(\overline{\Omega})$ and then defining

$$w_\varepsilon(x) := \alpha(x) + \varepsilon\left(\frac{R^2 - |x|^2}{2}\right), \quad x \in \overline{\Omega}, \qquad (10.32)$$

with $\varepsilon > 0$ and $R > 0$ chosen large enough so that $\overline{\Omega} \subset B_R(0)$, that is, $R^2 - |x|^2 > 0$ on $\overline{\Omega}$. One computes that

$$w_\varepsilon(x) > 0 \text{ for every } x \in \overline{\Omega} \quad \text{and} \quad D^2 w_\varepsilon(x) = -\varepsilon I < 0 \text{ for every } x,$$

so that (10.31) is satisfied. If, in addition, $u \le \alpha$ on $\partial\Omega$, then $u \le w_\varepsilon$ on $\partial\Omega$. Applying (10.29) yields

$$u \le w_\varepsilon \text{ on } \Omega,$$

with $\varepsilon > 0$ arbitrary. Taking the limit as $\varepsilon \to 0^+$ yields $u \le \alpha$ on Ω as desired and hence $u \in SA^+(X)$.

Step 2: $SA^+(X) \subset \{u \in USC(X) : u^+ \in SA(X)\}$. Suppose that $u \in SA^+(X)$. Given $\Omega \subset\subset X$ and $a \in \text{Aff}$ with $u^+ \le a$ on $\partial\Omega$, since $0 \le u^+$ it follows that $0 \le a$ on $\partial\Omega$. Now $-a$ is also affine, so it satisfies the maximum principle, and hence $0 \le a$ on $\overline{\Omega}$. This proves that $\alpha := a|_{\overline{\Omega}} \in \text{Aff}^+(\overline{\Omega})$. Since $u \in SA^+(X)$ and $u \le \alpha$ on $\partial\Omega$, it follows that $u \le \alpha$ on Ω. Also $0 \le \alpha = a$ on $\overline{\Omega}$. Therefore, $u^+ \le \alpha = a$ on $\overline{\Omega}$. This proves that $u^+ \in SA(X)$.

Step 3: $\{u \in USC(X) : u^+ \in \widetilde{\mathcal{P}}(X)\} \subset \widetilde{\mathcal{Q}}(X)$. We argue by contradiction to show that $u \notin \widetilde{\mathcal{Q}}(X) \Rightarrow u^+ \notin \widetilde{\mathcal{P}}(X)$. If $u \in USC(X) \setminus \widetilde{\mathcal{Q}}(X)$ then by the bad test jet lemma (Lemma 2.8) there exist $x_0 \in X$, $\varepsilon, \rho > 0$, $p \in \mathbb{R}^n$, and $(r, A) \notin \widetilde{\mathcal{Q}}$ such that

$$u(x) \le r + \langle p, x - x_0 \rangle + \tfrac{1}{2} \langle A(x - x_0), x - x_0 \rangle - \varepsilon |x - x_0|^2 \quad \forall x \in B_\rho(x_0) \tag{10.33}$$

and

$$u(x_0) = r. \tag{10.34}$$

Recall that $(r, A) \notin \widetilde{\mathcal{Q}}$ means that

$$r > 0 \quad \text{and} \quad A < 0. \tag{10.35}$$

Now $r > 0$ means that the right-hand side of (10.33) is positive for ρ sufficiently small. Therefore, (10.33) holds on $B_\rho(x_0)$ with $u(x)$ replaced by $u^+(x)$ (and we have equality at $x = x_0$). Thus u^+ has an upper test jet (r, p, A) at $x = x_0$ with $A < 0$, which proves that $u^+ \notin \widetilde{\mathcal{P}}(X)$.

Finally, for the sake of completeness, we include the following argument.

Step 4: $\widetilde{\mathcal{P}}(X) = SA(X)$. To see that $\widetilde{\mathcal{P}}(X) \subset SA(X)$, consider the simpler version $w_\varepsilon(x) := a(x) - \tfrac{\varepsilon}{2}|x|^2$ of (10.32), but with $a \in \text{Aff}$ arbitrary. Since $D^2 w_\varepsilon(x) = -\varepsilon I \notin \widetilde{\mathcal{P}}$ for every x, the subaffine property follows as in the argument after (10.32), but for any $u \in \widetilde{\mathcal{P}}(X)$, which proves that $u \in SA(X)$.

For the inclusion $SA(X) \subset \widetilde{\mathcal{P}}(X)$, see (10.33) with $A \notin \widetilde{\mathcal{P}}$, that is, $A < 0$ (negative definite). It follows that $u \notin SA(X)$. $\qquad\square$

Since $\tilde{\mathcal{Q}}(X) = \mathrm{SA}^+(X)$, we can restate the *subharmonic addition theorem* (10.25) in the following way by making use of Convention 3.11.

Theorem 10.8 (Subaffine-plus theorem). *If $\mathcal{F} \subset \mathcal{J}^2$ is a gradient-free subequation, then for any open set $X \subset \mathbb{R}^n$,*

$$\mathcal{F}(X) + \tilde{\mathcal{F}}(X) \subset \mathrm{SA}^+(X). \tag{10.36}$$

We conclude this section with a few remarks concerning Theorems 10.7 and 10.8.

Remark 10.9. It is easy to see that the set inclusion (10.25) is actually an equality. Thus (10.36) is also an equality.

Remark 10.10. The subaffine-plus Theorem 10.8 yields another proof of comparison (Theorem 10.3) for gradient-free subequations, because subaffine-plus functions clearly satisfy the (ZMP) as the function zero is subaffine plus.

Remark 10.11. While $\tilde{\mathcal{Q}}(X) = \mathrm{SA}^+(X) = \{u \in \mathrm{USC}(X) : u^+ \in \mathrm{SA}(X)\}$, one might wonder whether

$$\tilde{\mathcal{Q}}(X) = \{u \in \mathrm{USC}(X) : \forall \Omega \subset\subset X \text{ and } a \in \mathrm{Aff}, \text{ one has}$$
$$u \leq a^+ \text{ on } \partial\Omega \ \Rightarrow \ u \leq a^+ \text{ on } \Omega\}. \tag{10.37}$$

We leave it to the reader to show that the right-hand side of (10.37) is contained in $\tilde{\mathcal{Q}}(X) = \mathrm{SA}^+(X)$. Making use of an affine u, we now give an example that shows that (10.37) is *not* an equality. First, note that $\tilde{\mathcal{P}}(X) \subset \tilde{\mathcal{Q}}(X)$. Using the test (2.3) for C^2-functions, any $u \in \mathrm{Aff}(\mathbb{R}^n)$ belongs to $\tilde{\mathcal{P}}(\mathbb{R}^n)$ since $D^2 u(x) = 0 \in \tilde{\mathcal{P}}$ for all x. In dimension $n = 1$, consider

$$u(x) = x, \quad a(x) = 2(x - 1), \quad \text{and} \quad \Omega = (0, 2).$$

One has $u = a^+$ on $\partial\Omega$ but $u(1) = 1 > 0 = a^+(1)$.

10.3   FIRST ORDER AND PURE FIRST ORDER

By definition, a *first-order constraint set* is a subset $\mathcal{F} \subset \mathcal{J}^2 = \mathbb{R} \times \mathbb{R}^n \times \mathcal{S}(n)$ of the form

$$\mathcal{F} = \mathcal{F}' \times \mathcal{S}(n).$$

That is, the second-order factor is silent and the reduced constraint set $\mathcal{F}'$ is a subset of $\mathbb{R} \times \mathbb{R}^n$. Such a set $\mathcal{F}$ automatically satisfies the positivity condition (P). Hence, $\mathcal{F} = \mathcal{F}' \times \mathcal{S}(n)$ is a *first-order subequation* if $\mathcal{F}$ also satisfies the properties (N) and (T), which in terms of $\mathcal{F}'$ means

$$(r, p) \in \mathcal{F}' \;\Rightarrow\; (r + s, p) \in \mathcal{F}' \text{ for each } s \leq 0 \tag{10.38}$$

and

$$\mathcal{F}' = \overline{\mathrm{Int}\, \mathcal{F}'}. \tag{10.39}$$

Any monotonicity cone $\mathcal{M}$ for $\mathcal{F}$ can always be enlarged to include the silent factor $\mathcal{S}(n)$. Hence, we can replace our family of $(\gamma, \mathcal{D}, R)$-cones by the family of $(\gamma, \mathcal{D})$-*convex cones* whose elements are defined by

$$\mathcal{M}'(\gamma, \mathcal{D}) := \big\{ (r, p) \in \mathbb{R} \times \mathbb{R}^n : r \leq -\gamma |p| \text{ and } p \in \mathcal{D} \big\}, \tag{10.40}$$

where $\gamma \in [0, +\infty)$ and $\mathcal{D} \subset \mathbb{R}^n$ is a directional cone as in Definition 2.2. In particular,

$$\mathcal{M}(\gamma, \mathcal{D}, +\infty) = \mathcal{M}'(\gamma, \mathcal{D}) \times \mathcal{P} \subset \mathcal{M}'(\gamma, \mathcal{D}) \times \mathcal{S}(n),$$

so the strict approximators constructed in Theorem 6.3 for $\mathcal{M}(\gamma, \mathcal{D}, +\infty)$ are valid for $\mathcal{M} = \mathcal{M}'(\gamma, \mathcal{D}) \times \mathcal{S}(n)$ and for any domain $\Omega \subset\subset \mathbb{R}^n$. This proves that the (ZMP) holds for $\widetilde{\mathcal{M}}(\overline{\Omega})$-functions and hence we always have comparison in this case. The following notational device will be used for convenience: given a subequation constraint set $\mathcal{F}$ and a bounded domain Ω, denote

$$\mathcal{F}(\overline{\Omega}) := \big\{ u \in \mathrm{USC}(\overline{\Omega}) : u \text{ is } \mathcal{F}\text{-subharmonic on } \Omega \big\} = \mathrm{USC}(\overline{\Omega}) \cap \mathcal{F}(\Omega). \tag{10.41}$$

Theorem 10.12. *Let $\mathcal{F} = \mathcal{F}' \times \mathcal{S}(n) \subset \mathcal{J}^2$ be a first-order subequation such that $\mathcal{F}'$ is $\mathcal{M}'(\gamma, \mathcal{D})$-monotone with $\gamma \in [0, +\infty)$ and $\mathcal{D} \subset \mathbb{R}^n$ a directional cone which can be all of R^n. Then comparison always holds, that is, for any domain $\Omega \subset\subset \mathbb{R}^n$, given $u \in \mathcal{F}(\overline{\Omega})$ and $v \in \widetilde{\mathcal{F}}(\overline{\Omega})$, one has*

$$u + v \leq 0 \text{ on } \partial\Omega \;\Rightarrow\; u + v \leq 0 \text{ on } \Omega.$$

Finally, a *pure first-order subequation* $\mathcal{F} \subset \mathbb{R} \times \mathbb{R}^n \times \mathcal{S}(n)$ has both the first and third factors silent. That is,

$$\mathcal{F} = \mathbb{R} \times \mathcal{F}' \times \mathcal{S}(n)$$

with reduced constraint set $\mathcal{F}' \subset \mathbb{R}^n$ satisfying the single condition (T):

$$\mathcal{F}' = \overline{\mathrm{Int}\, \mathcal{F}'}. \tag{10.42}$$

The subequation conditions (N) and (P) are automatic in this case. However, for comparison $\mathcal{F}'$ must also satisfy the directionality property (D),

$$\mathcal{F}' + \mathcal{D} \subset \mathcal{F}', \quad \text{that is, } p \in \mathcal{F}' \implies (p+q) \in \mathcal{F}' \text{ for all } q \in \mathbb{R}^n \qquad (10.43)$$

for some nonempty closed convex cone $\mathcal{D} \subset \mathbb{R}^n$ with vertex at the origin. In this case, the comparison principle of Theorem 10.12 applies to $\mathcal{F}' \subset \mathbb{R}^n$ and arbitrary domains $\Omega \subset\subset \mathbb{R}^n$ by taking $\gamma = 0$.

10.4   ZERO-ORDER-FREE

By definition, a *zero-order-free constraint set* is a subset $\mathcal{F} \subset \mathcal{J}^2 = \mathbb{R} \times \mathbb{R}^n \times \mathcal{S}(n)$ of the form

$$\mathcal{F} = \mathbb{R} \times \mathcal{F}'.$$

That is, the zeroth-order factor is silent and the reduced constraint set $\mathcal{F}'$ is a subset of $\mathbb{R}^n \times \mathcal{S}(n)$. Such a set $\mathcal{F}$ automatically satisfies the negativity condition (N), but not the positivity condition (P) nor the topological condition (T). Here $\mathcal{F} = \mathbb{R} \times \mathcal{F}'$ is a *zero-order-free subequation* if the reduced constraint set $\mathcal{F}' \subset \mathbb{R}^n \times \mathcal{S}(n)$ satisfies

$$(p, A) \in \mathcal{F}' \implies (p, A+P) \in \mathcal{F}' \text{ for each } P \geq 0 \qquad (10.44)$$

and

$$\mathcal{F}' = \overline{\text{Int } \mathcal{F}'}. \qquad (10.45)$$

One might as well replace our family of $(\gamma, \mathcal{D}, R)$-cones by the family of cones whose elements are $\mathbb{R} \times \mathcal{M}'(\mathcal{D}, R)$, where

$$\mathcal{M}'(\mathcal{D}, R) := \left\{ (p, A) \in \mathbb{R}^n \times \mathcal{S}(n) : p \in \mathcal{D} \text{ and } A \geq \tfrac{|p|}{R} I \right\}, \qquad (10.46)$$

where $\mathcal{D} \subset \mathbb{R}^n$ is a nonempty closed convex cone with vertex at the origin. Notice that

$$\mathcal{M}(0, \mathcal{D}, R) \subset \mathbb{R} \times \mathcal{M}'(\mathcal{D}, R).$$

Since this inclusion is reversed by duality, our previous results apply, yielding the following comparison principle.

Theorem 10.13. *Let $\mathcal{F} = \mathbb{R} \times \mathcal{F}' \subset \mathcal{J}^2$ be a zero-order-free subequation such that $\mathcal{F}'$ is $\mathcal{M}'(\mathcal{D}, R)$-monotone. Then comparison holds; that is, given $u \in \mathcal{F}(\overline{\Omega})$ and $v \in \widetilde{\mathcal{F}}(\overline{\Omega})$, one has*

$$u + v \leq 0 \text{ on } \partial\Omega \implies u + v \leq 0 \text{ on } \Omega,$$

where Ω is any domain $\Omega \subset\subset \mathbb{R}^n$ contained in a translate of the truncated cone $\mathcal{D} \cap B_R(0)$ if $R < +\infty$ and Ω is an arbitrary bounded domain in the case $R = +\infty$.

10.5   SUMMARY

In this section we give a brief summary of the cases discussed above.

Summary Remark 10.14. For a subequation $\mathcal{F} \subset \mathcal{J}^2 = \mathbb{R} \times \mathbb{R}^n \times \mathcal{S}(n)$, there are six cases where a reduced constraint set $\mathcal{F}'$ can replace $\mathcal{F}$. In three of these cases, two factors of $\mathcal{F}$ are silent:

(1) (Pure second order). $\mathbb{R} \times \mathbb{R}^n$ is silent and $\mathcal{F} = \mathbb{R} \times \mathbb{R}^n \times \mathcal{F}'$ with $\mathcal{F}' \subsetneq \mathcal{S}(n)$ a closed set satisfying (P): $\mathcal{F}' + \mathcal{P} \subset \mathcal{F}'$.
(2) (Pure first order). $\mathbb{R} \times \mathcal{S}(n)$ is silent and $\mathcal{F} = \mathbb{R} \times \mathcal{F}' \times \mathcal{S}(n)$ with $\mathcal{F}' \subsetneq \mathbb{R}^n$ satisfying (T): $\mathcal{F}' = \overline{\text{Int}\,\mathcal{F}'}$.
(3) (Zeroth order). $\mathbb{R}^n \times \mathcal{S}(n)$ is silent and $\mathcal{F} = (-\infty, r_0] \times \mathbb{R}^n \times \mathcal{S}(n)$ with $r_0 \in \mathbb{R}$.

The remaining three cases have just one silent factor:

(4) (Zero-order-free). $\mathbb{R}$ is silent and $\mathcal{F} = \mathbb{R} \times \mathcal{F}'$ with $\mathcal{F}' \subsetneq \mathbb{R}^n \times \mathcal{S}(n)$ satisfying (P): $\mathcal{F}' + (\{0\} \times \mathcal{P}) \subset \mathcal{F}'$ and satisfying (T): $\mathcal{F}' = \overline{\text{Int}\,\mathcal{F}'}$.
(5) (Gradient-free). $\mathbb{R}^n$ is silent and $\mathcal{F} = \{(r, p, A) \in \mathcal{J}^2 : p \in \mathbb{R}^n \text{ and } (r, A) \in \mathcal{F}'\}$ with $\mathcal{F}' \subsetneq \mathbb{R} \times \mathcal{S}(n)$ a closed set satisfying (P) and (N): $\mathcal{F}' + (\mathcal{N} \times \mathcal{P}) \subset \mathcal{F}'$.
(6) (First order). $\mathcal{S}(n)$ is silent and $\mathcal{F} = \mathcal{F}' \times \mathcal{S}(n)$ with $\mathcal{F}' \subsetneq \mathbb{R} \times \mathbb{R}^n$ satisfying (N): $\mathcal{F}' + (\mathcal{N} \times \{0\}) \subset \mathcal{F}'$ and satisfying (T): $\mathcal{F}' = \overline{\text{Int}\,\mathcal{F}'}$.

Our main result Theorem 7.6 applies to case (5) (gradient-free) and hence to case (1) (pure second order) and to the trivial case (3) (zeroth order) with no further restrictions on the subequation, and yields comparison for arbitrary domains $\Omega \subset\subset \mathbb{R}^n$. A further condition on the reduced constraint set $\mathcal{F}'$ is required in each remaining case, in which the gradient constraint is *not* silent.

In case (2) (pure first order), in order to apply Theorem 7.6 we must assume directionality (D): $\mathcal{F}' + \mathcal{D} \subset \mathcal{F}'$ for a proper cone $\mathcal{D} \subset \mathbb{R}^n$. Then Theorem 7.6 applies and comparison holds on arbitrary domains $\Omega \subset\subset \mathbb{R}^n$.

In case (4) (zero-order-free), in order to apply Theorem 7.6 we must assume that $\mathcal{F}' \subset \mathbb{R}^n \times \mathcal{S}(n)$ is $\mathcal{M}'$-monotone for

$$\mathcal{M}'(\mathcal{D}, R) := \left\{ (p, A) \in \mathbb{R}^n \times \mathcal{S}(n) : p \in \mathcal{D} \text{ and } A \geq \tfrac{|p|}{R} I \right\}$$

for some convex cone $\mathcal{D} \subsetneq \mathbb{R}^n$ and some R with $0 < R \leq +\infty$. Theorem 7.6 applies and comparison holds on arbitrary domains $\Omega \subset\subset \mathbb{R}^n$ if $R = +\infty$. Otherwise, comparison holds for domains Ω contained in a translate of the truncated cone $\mathcal{D}_R := \mathcal{D} \cap B_R(0)$.

In case (6) (first order), in order to apply Theorem 7.6 we must assume that $\mathcal{F}' \subset \mathbb{R} \times \mathbb{R}^n$ is $\mathcal{M}$-monotone for the monotonicity cone

$$\mathcal{M}'(\gamma, \mathcal{D}) := \left\{ (r, p) \in \mathbb{R} \times \mathbb{R}^n : p \in \mathcal{D} \text{ and } r \leq -\gamma|p| \right\}$$

for some $\gamma \in [0, +\infty)$ and directional cone $\mathcal{D}$. In that case, Theorem 7.6 applies and comparison holds on arbitrary domains $\Omega \subset\subset \mathbb{R}^n$.

Although in case (3) (zeroth order) our results apply and comparison holds for all such subequations on arbitrary domains $\Omega \subset\subset \mathbb{R}^n$, no explicit discussion was presented since no constraint is placed on derivatives. This is the case when $\mathcal{F}$ is $\mathcal{M}(\gamma, \mathcal{D})$-monotone with $\gamma = 0$ and $\mathcal{D} = \mathbb{R}^n$ (so that $\mathcal{M}(\gamma, \mathcal{D}) = \mathcal{M}(\mathcal{N})$) which is included in Theorem 7.5. However, the proof of comparison is trivial, as $\widetilde{\mathcal{F}} = (-\infty, -r_0] \times \mathbb{R}^n \times \mathcal{S}(n)$.

Part III

Marrying Potential Theory to Operator Theory via the Correspondence Principle

Chapter Eleven

The Correspondence Principle for Compatible Operator-Subequation Pairs

In this chapter we will discuss some key issues concerning applications of the potential-theoretic comparison principles to comparison principles for constant-coefficient nonlinear operators.

Attention is restricted to the constant-coefficient case. This discussion goes beyond what is already given in [60]. It is helpful to be guided by two types of examples from the pure second-order case: one where the operator F is defined and "elliptic" on the full jet space $\mathcal{J}^2$, and one where F must be restricted (constrained) to a proper subset of $\mathcal{J}^2$ in order to be "elliptic." This dichotomy is illustrated by the minimal eigenvalue operator

$$F(r, p, A) := \lambda_{\min}(A), \quad \text{which is increasing in } A \text{ on all of } \mathcal{S}(n) \qquad (11.1)$$

and the Monge–Ampère operator

$$F(r, p, A) := \det A, \quad \text{which is increasing in } A \text{ only on } \mathcal{P} \subsetneq \mathcal{S}(n). \qquad (11.2)$$

Ideally, one would like to start from an equation

$$F(u, Du, D^2 u) = 0 \qquad (11.3)$$

defined by a function $F \colon \operatorname{dom}(F) \subseteq \mathcal{J}^2 \to \mathbb{R}$ and determine when there exists a subequation constraint set $\mathcal{F} \subset \mathcal{J}^2$ so that $\mathcal{F}$-subharmonics and $\mathcal{F}$-superharmonics ($-\widetilde{\mathcal{F}}$-subharmonics) correspond to viscosity subsolutions and supersolutions (with admissibility constraints) to the PDE (11.3). A natural attempt would be to realize $\mathcal{F}$ in the form

$$\mathcal{F} := \{ J = (r, p, A) \in \operatorname{dom}(F) : F(J) \geq 0 \}, \qquad (11.4)$$

where one would also need

$$\partial \mathcal{F} := \{ J = (r, p, A) \in \operatorname{dom}(F) : F(J) = 0 \}. \qquad (11.5)$$

The minimal monotonicity properties (P) and (N) for $\mathcal{F}$ can be deduced from the familiar monotonicity in (r, A) for $F = F(r, p, A)$, which, in general, will not hold on all of $\mathrm{dom}(F) \subseteq \mathcal{J}^2$. This imposes an a priori *constraint* on admissible values of the 2-jets and one will first need to restrict F and take $\mathrm{dom}(F) \subsetneq \mathcal{J}^2$ to be the *effective domain* on which F is suitably monotone in (r, A). In this *constrained case*, one might as well redefine $\mathrm{dom}(F) = \mathcal{F}$ so that the effective domain is a subequation constraint set. Ensuring that $\mathcal{F}$ has the needed topological property (T) is a more delicate matter and will be discussed below. Finally, to complete the applications, one would want to try to establish the needed structural conditions on F which ensure that $\mathcal{F}$ is suitably $\mathcal{M}$-monotone to yield comparison.

Classes of examples and illustrations will be given in the following sections. For example, unconstrained case examples include *canonical operators* as discussed in Section 11.4, and constrained case examples include *Gårding–Dirichlet polynomials* as discussed in Section 11.6. These classes are representative but, of course, not exhaustive for the dichotomy noted above. In particular, the minimal eigenvalue operator (11.1) is a canonical operator for the pure second-order subequation $\mathcal{P} \subset \mathcal{S}(n)$, and the Monge–Ampère operator (11.2) is one of the most basic and important Gårding–Dirichlet polynomials.

11.1   COMPATIBLE OPERATOR–SUBEQUATION PAIRS AND TOPOLOGICAL TAMENESS

We proceed with a precise discussion of the relationship between operators F and subequation constraint sets $\mathcal{F}$, beginning with the following definition of *compatibility* related to the desire to realize (11.5) in the constrained case.

Definition 11.1. A *compatible operator–subequation pair* $(F, \mathcal{F})$ consists of either

$$\mathcal{F} = \mathcal{J}^2 \text{ and } F \in C(\mathcal{J}^2) \quad \text{(the } \textit{unconstrained case}) \tag{11.6}$$

or

$$\text{a subequation } \mathcal{F} \subsetneq \mathcal{J}^2 \text{ and } F \in C(\mathcal{F}) \quad \text{(the } \textit{constrained case}). \tag{11.7}$$

In case (11.7), one requires that F and $\mathcal{F}$ are *compatible* in the following sense:

$$c_0 := \inf_{\mathcal{F}} F \text{ is finite} \tag{11.8}$$

and

$$\partial \mathcal{F} = \{ J \in \mathcal{F} : F(J) = c_0 \}. \tag{11.9}$$

Note that one can replace F by $F - c_0$ and reduce to the situation in which $c_0 = 0$.

Perhaps the simplest examples of compatible pairs come from the prototype operators noted above; namely, $(\lambda_{\min}(A), \mathcal{J}^2)$ and $(\det A, \mathbb{R} \times \mathbb{R}^n \times \mathcal{P})$ are compatible operator–subequation pairs in the unconstrained and constrained cases respectively. Notice that in this pure second-order case, we will also refer to

$$(\lambda_{\min}(A), \mathcal{S}(n)) \quad \text{and} \quad (\det A, \mathcal{P}) \tag{11.10}$$

as *compatible (pure second-order) pairs*, where one makes the obvious modification of Definition 11.1 in this and other reduced cases.

Compatibility of a pair $(F, \mathcal{F})$ will be used to define $\mathcal{F}$-*admissible viscosity subsolutions, supersolutions, and solutions* of the equation $F(u, Du, D^2u) = c$ for each *admissible level* $c \in F(\mathcal{F})$ (see Definition 11.12). Our treatment of comparison will be based on the *correspondence principle* of Theorem 11.13. For this principle, in addition to compatibility of the pair $(F, \mathcal{F})$, we also require that the pair has the minimal monotonicity of being *proper elliptic* ($\mathcal{M}_0 = \mathcal{N} \times \{0\} \times \mathcal{P}$-monotonicity) and a nondegeneracy property of *topological tameness* for F on $\mathcal{F}$. We proceed to discuss these two additional ingredients.

For a pair $\mathcal{F}$ and $F \in C(\mathcal{F}, \mathbb{R})$, the monotonicity of the operator F is related to the monotonicity of the set $\mathcal{F}$ as follows.

Definition 11.2. Suppose that $(F, \mathcal{F})$ is an operator–subequation pair. For any subset $\mathcal{M} \subset \mathcal{J}^2$, we will say that $(F, \mathcal{F})$ is $\mathcal{M}$-*monotone* if

$$\mathcal{F} \text{ is } \mathcal{M}\text{-monotone,} \quad \text{that is,} \quad \mathcal{F} + \mathcal{M} \subset \mathcal{F}, \tag{11.11}$$

and, in addition, F is $\mathcal{M}$-monotone on $\mathcal{F}$, that is,

$$F(J + J') \geq F(J) \quad \forall J \in \mathcal{F}, \ \forall J' \in \mathcal{M}. \tag{11.12}$$

Of course, if the pair $(F, \mathcal{F})$ is $\mathcal{M}$-monotone, then each particular upper level set such as $\mathcal{F}_0 := \{J \in \mathcal{F} : F(J) \geq 0\}$ will be $\mathcal{M}$-monotone, but the converse can fail to be true. However, by considering upper level sets for every $c \in \mathbb{R}$, the converse is trivially true.

Lemma 11.3. *Given an operator–subequation pair $(F, \mathcal{F})$ and a subset $\mathcal{M} \subset \mathcal{J}^2$, the pair $(F, \mathcal{F})$ is $\mathcal{M}$-monotone if and only if the upper level sets $\mathcal{F}_c := \{J \in \mathcal{F} : F(J) \geq c\}$ are $\mathcal{M}$-monotone for all $c \in \mathbb{R}$.*

Proof. The forward implication is a direct consequence of the defining conditions (11.11) and (11.12) of $\mathcal{M}$-monotonicity for $(F, \mathcal{F})$. For the converse, note that each $J \in \mathcal{F}$ belongs to $\mathcal{F}_c$ for each $c \leq F(J)$. Using the $\mathcal{M}$-monotonicity of every sublevel set $\mathcal{F}_c$, for each $J \in \mathcal{F}$ and each $J' \in \mathcal{M}$, one has

$$J + J' \in \mathcal{F}_c \subset \mathcal{F} \quad \text{for each } c \leq F(J),$$

which proves (11.11). Moreover, for each $J \in \mathcal{F}$ one has

$$F(J + J') \geq c \quad \text{for every } J' \in \mathcal{M} \text{ and every } c \leq F(J),$$

which for $c = F(J)$ proves (11.12). $\square$

Remark 11.4 (Admissible levels of F for compatible pairs). For a compatible pair $(F, \mathcal{F})$, when considering upper level sets $\mathcal{F}_c := \{J \in \mathcal{F} : F(J) \geq c\}$, lower level sets $\mathcal{F}^c := \{J \in \mathcal{F} : F(J) \leq c\}$, or level sets $\mathcal{F}(c) := \{J \in \mathcal{F} : F(J) = c\}$, one should of course restrict attention to values $c \in F(\mathcal{F})$ (those values which lie in the range of F). Otherwise, many statements become either redundant or empty (such as considering the $\mathcal{M}$-monotonicity of $\mathcal{F}_c$ in Lemma 11.3) if c is not in the range of F. In particular, in the constrained case, we will consider only those $c \in \mathbb{R}$ with $c \geq c_0 := \inf_{\mathcal{F}} F$.

Definition 11.5. Given a compatible operator–subequation pair $(F, \mathcal{F})$, a number $c \in \mathbb{R}$ is called an *admissible level* (for $(F, \mathcal{F})$) if $c \in F(\mathcal{F})$.

We can interpret operator *ellipticity* and *properness* using this lemma.

Definition 11.6. An operator–subequation pair $(F, \mathcal{F})$ is said to be *proper elliptic* if the pair $(F, \mathcal{F})$ is $\mathcal{M}_0 = \mathcal{N} \times \{0\} \times \mathcal{P}$-monotone in the sense of Definition 11.2.

Remark 11.7. For certain cases when F and $\mathcal{F}$ depend on the jet variable r, it is occasionally interesting to drop the requirement of negativity (N). Such is the case for questions concerning *generalized principal eigenvalues* (see [7, 10] and the references therein). In such cases, one simply requires that $\mathcal{F}$ satisfy properties (P) and (T) and that F be $\mathcal{M} = \{0\} \times \{0\} \times \mathcal{P}$-monotone on $\mathcal{F}$. In this case, we say that $(F, \mathcal{F})$ is *(degenerate) elliptic*. Note that in the reduced zero-order-free case, elliptic is the same as proper elliptic since the negativity (N) is automatic.

Corollary 11.8. *An operator–subequation pair $(F, \mathcal{F})$ is proper elliptic if and only if the upper level sets $\mathcal{F}_c := \{J \in \mathcal{F} : F(J) \geq 0\}$ are $\mathcal{M}_0 = \mathcal{N} \times \{0\} \times \mathcal{P}$-monotone for all admissible levels $c \in F(\mathcal{F})$.*

In the rest of this chapter, we will consider only compatible proper elliptic operator–subequation pairs in the sense of Definitions 11.1 and 11.2. In light of Corollary 11.8, for every admissible level $c \in F(\mathcal{F})$, the upper level set

$$\mathcal{F}_c := \{J \in \mathcal{F} : F(J) \geq c\} \text{ is closed, nonempty, and satisfies (P), (N).} \quad (11.13)$$

This means that each $\mathcal{F}_c$ is almost a subequation. One needs only the topological property (T). For this purpose, we place an additional structural condition on

F, which is easy to verify; for example, it is obviously satisfied if the operator F is real analytic.

Definition 11.9. A proper elliptic operator $F \in C(\mathcal{F}, \mathbb{R})$ is said to be *topologically tame* if the level set

$$\mathcal{F}(c) := \{J \in \mathcal{F} : F(J) = c\} \tag{11.14}$$

has empty interior for every admissible level $c \in F(\mathcal{F})$.

This condition rules out obvious pathologies. For example, if v is a local C^2-solution near $x_0 \in \mathbb{R}^n$ to $F(v, Dv, D^2v) = c$ with $J_{x_0}^2 v \in \operatorname{Int} \mathcal{F}(c) \neq \emptyset$, then for all C^2-functions φ with sufficiently small C^2-norm, $u := v + \varphi$ is also a local solution to $F(v, Dv, D^2v) = c$. Moreover, such a v always exists in this pathological case. For example, pick a 2-jet $J \in \operatorname{Int} \mathcal{F}(c)$ and let φ be the quadratic polynomial with 2-jet J at x_0. Then, picking φ with compact support near x_0 and small C^2-norm, one has lots of counterexamples to comparison.

Some strict monotonicity for the operator F provides a convenient way to rule out such pathologies, which we now discuss. As in (11.13) and (11.14), we will denote $\mathcal{F}_c := \{J \in \mathcal{F} : F(J) \geq c\}$ and $\mathcal{F}(c) := \{J \in \mathcal{F} : F(J) = c\}$.

Theorem 11.10 (Topological tameness equivalences). *Suppose that $(F, \mathcal{F})$ is a compatible proper elliptic operator–subequation pair which is $\mathcal{M}$-monotone for some convex cone subequation $\mathcal{M}$. Then the following are equivalent:*

(1) *F is topologically tame; that is, for each admissible level $c \in F(\mathcal{F})$, the level set $\mathcal{F}(c)$ has no interior.*
(2) *$F(J + J_0) > F(J)$ for each $J \in \mathcal{F}$ and each $J_0 \in \operatorname{Int} \mathcal{M}$.*
(3) *For some $J_0 \in \operatorname{Int} \mathcal{M}$, $F(J + tJ_0) > F(J)$ for each $J \in \mathcal{F}$ and for each $t > 0$.*
(4) *$\{J \in \mathcal{F} : F(J) > c\} = \operatorname{Int} \mathcal{F}_c$ for each admissible level $c \in F(\mathcal{F})$.*
(5) *$\mathcal{F}(c) = \mathcal{F}_c \cap (-\tilde{\mathcal{F}}_c)$ for each admissible level $c \in F(\mathcal{F})$.*

Before presenting the proof, some remarks are in order. Condition (2) is a strict version of the hypothesis that the operator F is $\mathcal{M}$-monotone. Condition (3) says that it is enough to have this strictness along the rays determined by a single $J_0 \in \operatorname{Int} \mathcal{M}$. Condition (4), by making use of the definitions, is the statement that *for every $c \in F(\mathcal{F})$ and every lower-semicontinuous function w,*

(4') *w is a supersolution of $F = c$ $\Leftrightarrow$ $-w$ is $\tilde{\mathcal{F}}_c$-subharmonic,* as will be made precise in Definition 11.12 and Theorem 11.13 below.

Proof of Theorem 11.10. There are seven implications to check. The proof that (1), (2), and (3) are equivalent does not use the assumption that F and $\mathcal{F}$ are compatible.

(2) $\Rightarrow$ (3). This is obvious.

$(3) \Rightarrow (1)$. Assume that (1) is false, that is, for some $c \in \mathbb{R}$ there exists $J \in \mathrm{Int}\, \mathcal{F}(c)$. Then given $J_0 \in \mathrm{Int}\, \mathcal{M}$ with $J_0 \neq 0$, one has $J + t J_0 \in \mathrm{Int}\, \mathcal{F}(c)$ for each $t > 0$ sufficiently small, which contradicts (3).

$(1) \Rightarrow (2)$. Assume that (2) is false, that is, there exist $J_1 \in \mathcal{F}$ and $J_0 \in \mathrm{Int}\, \mathcal{M}$ such that $F(J_1 + J_0) \leq F(J_1)$. Since F is $\mathcal{M}$-monotone, one has $F(J_1 + J_0) \geq F(J_1)$ and hence

$$F(J_1 + J_0) = F(J_1) := c \quad \text{for some } J_1 \in \mathcal{F} \text{ and for some } J_0 \in \mathrm{Int}\, \mathcal{M}. \quad (11.15)$$

We will show that for $\varepsilon > 0$ sufficiently small, $J_1 + (1 - \varepsilon) J_0 \in \mathrm{Int}\, \mathcal{F}(c)$, which contradicts (1).

First, note that

$$\text{if } J \in \mathcal{U}_1 := J_1 + \mathrm{Int}\, \mathcal{M} \text{ then } J \in \mathcal{F} \text{ and } c := F(J_1) \leq F(J) \quad (11.16)$$

by the $\mathcal{M}$-monotonicity of the operator F.

Second, since $J_0 \in \mathrm{Int}\, \mathcal{M}$, $J_0 - \mathcal{B} \subset \mathcal{M}$ if $\mathcal{B}$ is a small ball about the origin in $\mathcal{J}^2$. Next we prove that

$$\text{if } J \in \mathcal{U}_2 := J_1 + J_0 - (\mathcal{B} \cap \mathrm{Int}\, \mathcal{M}) \text{ then } J \in \mathcal{F} \text{ and } F(J) = c. \quad (11.17)$$

Suppose that $J := J_1 + J_0 - J'$ with $J' \in (\mathcal{B} \cap \mathrm{Int}\, \mathcal{M})$. Then $J \in \mathcal{F}$ since $J_1 \in \mathcal{F}$ and $J_0 - J' \in \mathcal{M}$. Moreover, since $J' \in \mathcal{M}$ one has

$$F(J) \leq F(J + J') = F(J_1 + J_0) = c,$$

again using that F is $\mathcal{M}$-monotone. Hence, the open set

$$\mathcal{U}_1 \cap \mathcal{U}_2 \subset \mathrm{Int}\, \mathcal{F}(c). \quad (11.18)$$

Finally, $J_1 + (1 - \varepsilon) J_0 \in \mathcal{U}_1$ if $0 < \varepsilon < 1$ and $J_1 + (1 - \varepsilon) J_0 \in \mathcal{U}_2$ if ε is small enough to ensure $\varepsilon J_0 \in \mathcal{B}$, proving that $\mathcal{U}_1 \cap \mathcal{U}_2 \neq \emptyset$.

$(2) \Rightarrow (4)$. The compatibility assumption (11.9) of Definition 11.1 can be rephrased, since $\mathcal{F} = \mathcal{F}_{c_0}$, as $\partial \mathcal{F} := \mathcal{F} \setminus \mathrm{Int}\, \mathcal{F} = \{J \in \mathcal{F} : F(J) = c_0\}$, that is,

$$\mathrm{Int}\, \mathcal{F} = \{J \in \mathcal{F} : F(J) > c_0\}. \quad (11.19)$$

Now, by (11.19), for $c \geq c_0$,

$$\{F > c\} := \{J \in \mathcal{F} : F(J) > c\} = \{J \in \mathrm{Int}\, \mathcal{F} : F(J) > c\} \quad (11.20)$$

is an open subset of $\mathcal{J}^2$. Since $\{F > c\}$ is contained in $\mathcal{F}_c$, it is part of the interior of $\mathcal{F}_c$.

Conversely, suppose that $J \in \text{Int}\,\mathcal{F}_c$. Pick $J_0 \in \text{Int}\,\mathcal{M}$. Then for $\varepsilon > 0$ sufficiently small, $J - \varepsilon J_0 \in \mathcal{F}_c$. By (2) one has

$$F(J) = F((J - \varepsilon J_0) + \varepsilon J_0) > F(J - \varepsilon J_0) \geq c,$$

and hence $J \in \{F > c\}$.

(4) $\Rightarrow$ (1). If (1) were false, then for some $c \in \mathbb{R}$ there would exist an open set $\mathcal{U} \subset \mathcal{F}(c) \subset \mathcal{F}_c$. Thus $\mathcal{U} \subset \text{Int}\,\mathcal{F}_c$ but $\mathcal{U} \not\subset \{J \in \mathcal{F} : F(J) > c\}$, so that (4) would be false.

(4) $\Rightarrow$ (5). One has

$$\mathcal{F}_c \cap (-\widetilde{\mathcal{F}}_c) = \{J \in \mathcal{F} : F(J) \geq c\} \cap (\sim \text{Int}\,\mathcal{F}_c),$$

which, by (4), equals $\{J \in \mathcal{F} : F(J) \geq c \text{ and } F(J) \leq c\} := \mathcal{F}(c)$.

(5) $\Rightarrow$ (1). Suppose that (1) is false. Then, for some $c \in \mathbb{R}$, there is an open set $\mathcal{U} \subset \mathcal{F}(c)$. Hence, $\mathcal{U} \subset \text{Int}\,\mathcal{F}_c$ so that $\mathcal{F}(c) \subset \sim(\text{Int}\,\mathcal{F}_c)$ is false, which contradicts (5). $\qquad\square$

Remark 11.11. Under the hypotheses of the theorem, if F is also topologically tame, then it follows easily from (4) that for each $c \in F(\mathcal{F})$, the upper level set $\mathcal{F}_c$ is a subequation.

11.2   THE CORRESPONDENCE PRINCIPLE FOR COMPATIBLE PAIRS

We now discuss an important consequence of Theorem 11.10 which will be essential for our treatment of comparison for classes of nonlinear operators in the next chapter. Recall that φ is a C^2 *(upper/lower) test function for u at x_0* if

$$u - \varphi \gtrless 0 \text{ near } x_0 \quad \text{and} \quad u - \varphi = 0 \text{ at } x_0.$$

We will denote by $J^{2,\pm}_{x_0} u \subset \mathcal{J}^2$ the spaces of *(upper/lower) test jets for u at x_0*, that is, the set of all $J = J^2_{x_0} \varphi$ where φ is a C^2 (upper/lower) test function for u at x_0.

Definition 11.12. Let $(F, \mathcal{F})$ be a compatible operator–subequation pair as in Definition 11.1. Let Ω be a domain in $\mathbb{R}^n$ and let $c \in F(\mathcal{F})$ be an admissible level.

(a) A function $u \in \text{USC}(\Omega)$ is said to be an *$\mathcal{F}$-admissible viscosity subsolution of $F(u, Du, D^2u) = c$ in Ω* if, for every $x_0 \in \Omega$, one has

$$J \in J^{2,+}_{x_0} u \;\Rightarrow\; J \in \mathcal{F} \text{ and } F(J) \geq c. \tag{11.21}$$

(b) A function $u \in \mathrm{LSC}(\Omega)$ is said to be an $\mathcal{F}$-*admissible viscosity supersolution* of $F(u, Du, D^2u) = c$ *in* Ω if

$$J \in J^{2,-}_{x_0} u \;\Rightarrow\; \text{either } [J \in \mathcal{F} \text{ and } F(J) \le c] \text{ or } J \notin \mathcal{F}. \tag{11.22}$$

We say that $u \in C(\Omega)$ is an $\mathcal{F}$-*admissible viscosity solution* of $F(u, Du, D^2u) = c$ *in* Ω if both (a) and (b) hold.

Note that in the unconstrained case $(\mathcal{F} = \mathcal{J}^2)$ these definitions of $\mathcal{J}^2$-admissible viscosity (sub, super) solutions are standard and called merely *viscosity (sub, super) solutions*, respectively. In the constrained case $(\mathcal{F} \subsetneq \mathcal{J}^2)$, we are taking a systematic approach to what is often done ad hoc for particular examples.

The following result formalizes the previous considerations in order to illustrate a general situation in which the potential-theoretic approach in terms of subequation constraint sets $\mathcal{F}_c = \{J \in \mathcal{F} : F(J) \ge c\}$ corresponds to the PDE approach of $\mathcal{F}$-admissible viscosity solutions to $F = c$.

Theorem 11.13 (Correspondence principle for compatible pairs). *Suppose that* $(F, \mathcal{F})$ *is a compatible proper elliptic operator–subequation pair which is* $\mathcal{M}$-*monotone for some convex cone subequation* $\mathcal{M}$. *Suppose also that* F *is topologically tame. Then, for every admissible level* $c \in F(\mathcal{F})$ *and for every domain* $\Omega \subset \mathbb{R}^n$, *one has*

(a) $u \in \mathrm{USC}(\Omega)$ *is an* $\mathcal{F}$-*admissible viscosity subsolution of* $F(u, Du, D^2u) = c$ *in* Ω *if and only if* u *is* $\mathcal{F}_c$-*subharmonic on* Ω;
(b) $u \in \mathrm{LSC}(\Omega)$ *is an* $\mathcal{F}$-*admissible viscosity supersolution of* $F(u, Du, D^2u) = c$ *in* Ω *if and only if* u *is* $\mathcal{F}_c$-*superharmonic on* Ω, *which is, by Definition 3.6, saying that* $-u$ *is* $\widetilde{\mathcal{F}}_c$-*subharmonic on* Ω.

In particular, for every admissible level $c \in F(\mathcal{F})$, *one has comparison for the subequation* $\mathcal{F}_c$ *on a domain* Ω *if and only if one has comparison for the equation* $F(u, Du, D^2u) = c$ *on* Ω.

We recall that by comparison we mean the validity of the comparison principle

$$u \le w \text{ on } \partial\Omega \;\Rightarrow\; u \le w \text{ on } \Omega \tag{11.23}$$

for all pairs $u \in \mathrm{USC}(\overline{\Omega})$ and $w \in \mathrm{LSC}(\overline{\Omega})$ which satisfy (a) and (b) respectively.

Proof of Theorem 11.13. For part (a), definition (11.22) of $u \in \mathrm{USC}(\Omega)$ being an $\mathcal{F}$-admissible subsolution in x_0 is equivalent to the statement that

$$J^{2,+}_{x_0} u \subset \{J \in \mathcal{F} : F(J) \ge c\} := \mathcal{F}_c, \tag{11.24}$$

where $J^{2,+}_{x_0} u \subset \mathcal{F}_c$ defines $\mathcal{F}_c$-subharmonicity in x_0.

For part (b), definition (11.22) of $u \in \mathrm{LSC}(\Omega)$ being an $\mathcal{F}$-admissible super-solution in x_0 is equivalent to the statement that

$$J^{2,-}_{x_0} u \subset \{J \in \mathcal{F} : F(J) \leq c\} \cup (\sim \mathcal{F}). \tag{11.25}$$

Since F is topologically tame and $\mathcal{M}$-monotone for some convex cone subequation $\mathcal{M}$, Theorem 11.10(4) yields

$$\mathrm{Int}\, \mathcal{F}_c = \{J \in \mathcal{F} : F(J) > c\}, \tag{11.26}$$

and hence

$$\sim \mathrm{Int}\, \mathcal{F}_c = \{J \in \mathcal{F} : F(J) \leq c\} \cup (\sim \mathcal{F}). \tag{11.27}$$

Combining (11.25) and (11.27) yields

$$J^{2,-}_{x_0} u \subset\ \sim \mathrm{Int}\, \mathcal{F}_c,$$

which by formula (3.10) of Remark 3.7 is one way to define that $u \in \mathrm{LSC}(\Omega)$ is $\mathcal{F}_c$-superharmonic. $\qquad\square$

In the following Sections 11.3, 11.4, and 11.5, we will discuss how the monotonicity property $\mathcal{F} + \mathcal{M} \subset \mathcal{F}$ provides additional structure, with important consequences for any subequation $\mathcal{F}$ which admits a monotonicity subequation cone $\mathcal{M}$. This *structure theorem* (see Theorem 11.14) provides the existence of a *canonical operator* $F \in C(\mathcal{J}^2)$ associated to any such $\mathcal{M}$-monotone subequation $\mathcal{F}$. Such operators give rise to a rich class of examples for which the correspondence principle of Theorem 11.13 applies. Moreover, the structure theorem yields a uniqueness result for *subequation branches* of a given equation $\mathcal{H} \subset \mathcal{J}^2$ (see Corollary 11.15 and Proposition 11.37). The beautiful class of *Gårding polynomial operators* is discussed in Section 11.6. This will further expand (in Chapter 12) applications of the correspondence principle for obtaining additional comparison principles for compatible operator–subequation pairs (in both constrained and unconstrained cases).

11.3 A STRUCTURE THEOREM DERIVED FROM SUBEQUATION MONOTONICITY

The family of lines in $\mathcal{J}^2$ in any fixed direction $J_0 \in \mathrm{Int}\,\mathcal{M}$ provides structure to a subequation $\mathcal{F}$ which admits the monotonicity cone subequation $\mathcal{M}$. We recall that part of the definition of $\mathcal{F} \subset \mathcal{J}^2$ being a subequation is that $\mathcal{F} \neq \emptyset, \mathcal{J}^2$ and part of the definition of $\mathcal{M}$ being a monotonicity cone subequation is that $\mathcal{M}$ has nonempty interior. The following fundamental result is contained in [49,

Lemma 9.9, part (2)], which was stated for manifolds but not proven there. See also [48, Theorem 3.2] for the construction in the pure second-order case.

Theorem 11.14 (Structure theorem). *Suppose that $\mathcal{F} \subset \mathcal{J}^2$ is a subequation constraint set which admits a monotonicity cone subequation $\mathcal{M}$. Fix $J_0 \in \text{Int } \mathcal{M}$. Given $J \in \mathcal{J}^2$ arbitrary, the set*

$$I_J := \{t \in \mathbb{R} : J + tJ_0 \in \mathcal{F}\} \tag{11.28}$$

is a closed interval of the form $[t_J, +\infty)$ with $t_J \in \mathbb{R}$ (finite). Moreover,

(a) $J + tJ_0 \notin \mathcal{F} \Leftrightarrow t < t_J;$
(b) $J + t_J J_0 \in \partial \mathcal{F};$
(c) $J + tJ_0 \in \text{Int } \mathcal{F} \Leftrightarrow t > t_J;$

and any one of the relations (a), (b), *or* (c) *uniquely determines $t_J \in \mathbb{R}$ from $J \in \mathcal{J}^2$ and $J_0 \in \text{Int } \mathcal{M}$ in the sense that*

$$t_J \text{ is the unique element of } \mathbb{R} \text{ for which } J + t_J J_0 \in \partial \mathcal{F} \tag{11.29}$$

and

$$t_J = \inf\{t \in \mathbb{R} : J + tJ_0 \in \text{Int } \mathcal{M}\} = \sup\{t \in \mathbb{R} : J + tJ_0 \notin \mathcal{F}\}. \tag{11.30}$$

Proof. With $J \in \mathcal{J}^2$ arbitrary, we first show that $I_J = [t_J, +\infty)$ for a unique value $t_J \in \mathbb{R}$. The proof involves four steps.

Step 1: One has $J + tJ_0 \in \mathcal{M}$ for all t sufficiently large. Indeed, since $\mathcal{M}$ is a cone, for $t > 0$ this is equivalent to having

$$\frac{1}{t} J + J_0 \in \mathcal{M} \quad \text{for all } t \text{ sufficiently large,}$$

which holds since $J_0 \in \text{Int } \mathcal{M}$.

Step 2: $I_J := \{t \in \mathbb{R} : J + tJ_0 \in \mathcal{F}\}$ is nonempty. Indeed, since $\mathcal{F} \neq \emptyset$ we can pick any $J_1 \in \mathcal{F}$ and then notice that

$$J + tJ_0 = J_1 + (J - J_1) + tJ_0 \in \mathcal{F} + \mathcal{M} \subset \mathcal{F} \quad \text{for all } t \text{ sufficiently large,}$$

since $\mathcal{F}$ is $\mathcal{M}$-monotone and $(J - J_1) + tJ_0 \in \mathcal{M}$ for all large t by step 1.

Step 3: If $t \in I_J$ then $t + s \in I_J$ for each $s \geq 0$. It is enough to notice that

$$J + (t + s)J_0 = (J + tJ_0) + sJ_0 \in \mathcal{F} + \mathcal{M} \subset \mathcal{F},$$

since $\mathcal{F}$ is $\mathcal{M}$-monotone.

By steps 2 and 3, one has either $I_J = [t_J, +\infty)$ or $I = \mathbb{R}$.

Step 4: One has $I_J \neq \mathbb{R}$ and hence $I_J = [t_J, +\infty)$ for some $t_J \in \mathbb{R}$. Suppose not. Then $J + tJ_0 \in \mathcal{F}$ for all $t \in \mathbb{R}$. Let $J' \in \mathcal{J}^2$ be arbitrary. By step 1, there exists $s \geq 0$ such that

$$(J' - J) + sJ_0 \in \mathcal{M}$$

and hence

$$(J + tJ_0) + (J' - J + sJ_0) \in \mathcal{F} + \mathcal{M} \subset \mathcal{F} \quad \text{for all } t \in \mathbb{R}. \tag{11.31}$$

Taking $t = -s$ in (11.31) yields $J' \in \mathcal{F}$ for arbitrary $J' \in \mathcal{F}$, which contradicts $\mathcal{F} \neq \mathcal{J}^2$.

It remains to verify claims (a), (b), and (c). Claim (a) follows from the fact that, by construction,

$$t_J = \min\{t \in \mathbb{R} : J + tJ_0 \in \mathcal{F}\}. \tag{11.32}$$

For claim (b), notice that $J + (t_J - \varepsilon)J_0 \notin \mathcal{F}$ for each $\varepsilon > 0$ and hence $J + t_J J_0 \notin \operatorname{Int} \mathcal{F}$. Therefore,

$$J + t_J J_0 \in \mathcal{F} \setminus \operatorname{Int} \mathcal{F} = \partial \mathcal{F}.$$

For claim (c), if $s > 0$ then

$$J + (t_J + s)J_0 = (J + t_J J_0) + sJ_0 \in \partial \mathcal{F} + \operatorname{Int} \mathcal{M} \subset \mathcal{F} + \operatorname{Int} \mathcal{M} \subset \operatorname{Int} \mathcal{F},$$

by the set identity (4.13) of Proposition 4.7. This proves the implication ($\Leftarrow$) of claim (c). However, by claims (a) and (b), one has

$$t \leq t_J \ \Rightarrow\ J + tJ_0 \in \partial \mathcal{F} \cup (\sim \mathcal{F}) = \sim \operatorname{Int} \mathcal{F},$$

which is contrapositive to the implication ($\Rightarrow$) of claim (c).

Finally, formulas (11.29) and (11.30) follow from (a), (b), and (c). $\qquad\square$

An important corollary of the structure theorem (Theorem 11.14) is that $\mathcal{M}$-monotone subequations $\mathcal{F}$ are uniquely determined by their boundaries $\partial \mathcal{F}$.

Corollary 11.15. *Suppose that $\mathcal{F} \subset \mathcal{J}^2$ is a subequation constraint set which admits a monotonicity cone subequation $\mathcal{M}$. Then*

$$\mathcal{F} = \partial \mathcal{F} + \mathcal{M}. \tag{11.33}$$

Proof. Since $\mathcal{F} + \mathcal{M} \subset \mathcal{F}$, we have $\partial \mathcal{F} + \mathcal{M} \subset \mathcal{F}$. For the reverse inclusion, fix any $J_0 \in \operatorname{Int} \mathcal{M}$. Given $J \in \mathcal{F}$, by the $\mathcal{M}$-monotonicity of $\mathcal{F}$ with respect to the cone $\mathcal{M}$, one has

$$J + tJ_0 \in \mathcal{F} \quad \text{for every } t \geq 0. \tag{11.34}$$

Since t_J is the minimal $t \in \mathbb{R}$ for which $J + tJ_0 \in \mathcal{F}$, one has $t_J \leq 0$ and hence

$$J = (J + t_J J_0) - t_J J_0 \in \partial \mathcal{F} + \mathcal{M}. \qquad \square$$

The structure theorem (Theorem 11.14) also provides a tool for showing the existence of *canonical operators*, as well as *graphing functions for boundaries* $\partial \mathcal{F}$ under the monotonicity assumption of this structure theorem.

11.4 CANONICAL OPERATORS FOR SUBEQUATIONS WITH MONOTONICITY

The structure theorem (Theorem 11.14) provides a canonical procedure for constructing an operator $F \in C(\mathcal{J}^2)$ with "nice" properties, associated to a given subequation $\mathcal{F}$ as long as $\mathcal{F}$ admits a monotonicity cone subequation $\mathcal{M}$ (and one fixes an element $J_0 \in \text{Int } \mathcal{M}$). First, $(F, \mathcal{F})$ is a compatible operator–subequation pair with monotonicity $\mathcal{M}$, providing $\mathcal{F}$ with at least one compatible operator which is natural. Second, F is defined on all of $\mathcal{J}^2$ and $(F, \mathcal{J}^2)$ will be shown to be a compatible (unconstrained case) proper elliptic operator–subequation pair in the sense of Definition 11.1, where the operator F is topologically tame on $\mathcal{J}^2$. This gives a rich family (including $(F, \mathcal{F})$) of pairs for which the correspondence principle of Theorem 11.13 holds. The canonical operator F is closely related to the potential theory equation $\partial \mathcal{F}$. For *any* hyperplane (codimension-one subspace) $W_0 \subset \mathcal{J}^2$ transverse to the line (one-dimensional subspace) $[J_0]$ through J_0, the restriction $g := F|_{W_0} : W_0 \to \mathbb{R}$ is the unique graphing function for $\partial \mathcal{F}$ over W_0, that is,

$$\partial \mathcal{F} = \{ J + g(J)J_0 : J \in W_0 \}.$$

A judicious choice of the hyperplane W_0 results in the Lipschitz regularity of g and F with respect to a natural seminorm, as will be discussed in the next section.

The canonical operator is defined in terms of the structure theorem as follows.

Definition 11.16. Suppose that $\mathcal{F} \subset \mathcal{J}^2$ is a subequation constraint set which admits a monotonicity cone subequation $\mathcal{M}$. For fixed $J_0 \in \text{Int } \mathcal{M}$, the *canonical operator for $\mathcal{F}$ (determined by J_0)* $F : \mathcal{J}^2 \to \mathbb{R}$ is defined by

$$F(J) := -t_J, \quad \text{where } t_J \text{ is defined by (11.29)} \tag{11.35}$$

(or by either of the two formulas in (11.30)).

We proceed to analyze the properties of the canonical operator outlined above, beginning with some structural properties.

Proposition 11.17 (Structural properties of the canonical operator). *Suppose that F is the canonical operator for $\mathcal{F}$ (determined by a fixed $J_0 \in \text{Int}\,\mathcal{M}$). Then*

(a) *F decomposes $\mathcal{J}^2$ into three disjoint pieces:*

$$\partial\mathcal{F} = \{F(J) = 0\}, \quad \text{Int}\,\mathcal{F} = \{F(J) > 0\}, \quad \text{and} \quad \mathcal{J}^2 \setminus \mathcal{F} = \{F(J) < 0\}; \quad (11.36)$$

(b) *F is strictly increasing in the direction J_0, in fact,*

$$F(J + tJ_0) = F(J) + t \quad \text{for each } J \in \mathcal{J}^2 \text{ and for each } t \in \mathbb{R}; \quad (11.37)$$

(c) *F is proper elliptic on $\mathcal{J}^2$, in fact, F is $\mathcal{M}$-monotone on $\mathcal{J}^2$; that is,*

$$F(J + J') \geq F(J) \quad \text{for each } J \in \mathcal{J}^2 \text{ and for each } J' \in \mathcal{M}. \quad (11.38)$$

Proof. Decomposition (11.36) of $\mathcal{J}^2$ is a restatement of (a), (b), and (c) of the structure theorem (Theorem 11.14), as follows. First, consider those $J \in \mathcal{J}^2$ such that $F(J) := -t_J = 0$. By Theorem 11.14, we have

$$J \in \partial\mathcal{F}, \quad J + tJ_0 \notin \mathcal{F} \text{ for all } t < 0 \quad \text{and} \quad J + tJ_0 \in \text{Int}\,\mathcal{F} \text{ for all } t > 0,$$

and hence $\partial\mathcal{F} = \{F(J) = 0\}$, as desired. For arbitrary $J \in \mathcal{J}^2$, use the definition of F and relations (a) and (c) of Theorem 11.14 to find

$$F(J) < 0 \iff t_J > 0 \iff J \notin \mathcal{F}$$

and

$$F(J) > 0 \iff t_J < 0 \iff J \in \text{Int}\,\mathcal{F}.$$

Next, using the definition of F, property (11.37) requires showing that for every $J \in \mathcal{J}^2$,

$$t_{(J + tJ_0)} = t_J - t \quad \text{for every } t \in \mathbb{R}.$$

By construction, t_J and $t_{(J + tJ_0)}$ are the unique real numbers such that

$$J + t_J J_0 \in \partial\mathcal{F} \quad \text{and} \quad (J + tJ_0) + t_{(J + tJ_0)} J_0 \in \partial\mathcal{F}$$

and hence $t_J = t + t_{(J + tJ_0)}$, proving (11.37).

Finally, we note that property (11.37) of part (b) yields

$$F(J + tJ_0) > F(J) \quad \text{for each } J \in \mathcal{J}^2 \text{ and for each } t > 0,$$

which is condition (3) of Theorem 11.10 concerning topological tameness. Hence, all of the other equivalent forms (1), (2), (4), and (5) hold, where the strict monotonicity condition (2) is stronger than condition (11.38). □

The following result shows that property (11.37) plus any one of the relations in (11.36) uniquely determine F (for $J_0 \in \text{Int}\,\mathcal{M}$ fixed) and hence they could be taken as defining properties for the canonical F determined by J_0.

Proposition 11.18. *Suppose that for some $J_0 \in \mathcal{J}^2$, an operator $F\colon \mathcal{J}^2 \to \mathbb{R}$ satisfies the affine property (11.37), that is,*

$$F(J + tJ_0) = F(J) + t \quad \text{for each } J \in \mathcal{J}^2 \text{ and for each } t \in \mathbb{R}. \tag{11.39}$$

Suppose that $\mathcal{F}$ is a subequation which admits a monotonicity cone subequation $\mathcal{M}$ with $J_0 \in \text{Int}\,\mathcal{M}$. If any one of the relations

(a) $\partial \mathcal{F} = \{F(J) = 0\}$, (b) $\text{Int}\,\mathcal{F} = \{F(J) > 0\}$, *or* (c) $\mathcal{J}^2 \setminus \mathcal{F} = \{F(J) < 0\}$

holds, then F is the canonical operator (determined by J_0) for $\mathcal{F}$.

Proof. With $J_0 \in \text{Int}\,\mathcal{M}$ fixed, the affine property (11.39) shows that for each $J \in \mathcal{J}^2$, the restriction of F to the line $\ell_J = \{J + tJ_0 : t \in \mathbb{R}\}$ is continuous, strictly increasing, and has range equal to $\mathbb{R}$. Hence, there is a unique value $t^* \in \mathbb{R}$ such that

$$0 = F(J + t^* J_0) = F(J) + t^*. \tag{11.40}$$

By Definition 11.16, we only need to show that $t^* = t_J$, where $t_J \in \mathbb{R}$ is the critical parameter in Theorem 11.14 which divides ℓ_J into the three pieces (a unique point on $\partial \mathcal{F}$, an open ray in $\text{Int}\,\mathcal{M}$, and an open ray in $\mathcal{J}^2 \setminus \mathcal{F}$). The three relations (a), (b), and (c) imply that $t^* = t_J$ as defined by (11.29) and the first and second formulas of (11.30) respectively. □

Next we observe that an immediate consequence of the structure theorem (Theorem 11.14) and the definition of the canonical operator F for $\mathcal{F}$ is that the *equation* $\partial \mathcal{F} = \mathcal{F} \cap (-\widetilde{\mathcal{F}})$ can be graphed and the canonical operator can be recovered from the graphing function g.

Proposition 11.19 (Canonical operators and graphing the equation $\partial \mathcal{F}$). *Suppose that $\mathcal{F} \subset \mathcal{J}^2$ is a subequation constraint set which admits a monotonicity cone subequation $\mathcal{M}$. Let F be the canonical operator for $\mathcal{F}$ determined by $J_0 \in \text{Int}\,\mathcal{M}$. Fix $W_0 \subset \mathcal{J}^2$ a hyperplane in $\mathcal{J}^2$ transversal to $[J_0]$. Then one has*

(a) *the equation $\partial \mathcal{F} \subset W_0 \oplus [J_0]$ is the graph of $g\colon W_0 \to \mathbb{R}$ defined by*

$$g(J') := -F(J'), \quad J' \in W_0, \tag{11.41}$$

which is to say that $g := -F|_{W_0}$, *or equivalently,*

$$\partial F = \{ J \in \mathcal{J}^2 : J = J' + g(J')J_0, \text{ where } J' \in W_0 \}; \tag{11.42}$$

(b) *the epigraph of g satisfies*

$$\mathcal{F} = \{ J \in \mathcal{J}^2 : J = J' + tJ_0, \text{ where } t \geq g(J') \text{ and } J' \in W_0 \}; \tag{11.43}$$

(c) *the canonical operator F is recovered from the graphing function g by*

$$F(J) = F(J' + tJ_0) = t - g(J') \quad \text{for each } J \in \mathcal{J}^2. \tag{11.44}$$

Proof. Consider the splitting $\mathcal{J}^2 = W_0 \oplus [J_0]$. Each $J \in \mathcal{J}^2$ can be decomposed uniquely into

$$J = J' + tJ_0 \quad \text{with } J' \in W_0 \text{ and } t \in \mathbb{R}. \tag{11.45}$$

With respect to this decomposition (11.45), the structure theorem (Theorem 11.14) with $J' \in W_0 \subset \mathcal{J}^2$ arbitrary says that

$$J' + tJ_0 \in \partial \mathcal{F} \iff t = t_{J'} = -F(J') = g(J')$$

and

$$J' + tJ_0 \in \partial \mathcal{F} \iff t \geq t_{J'} = -F(J') = g(J'),$$

which gives the claims in parts (a) and (b). Part (c) is immediate. $\qquad\square$

We summarize the above considerations by noting that canonical operators F associated to subequations $\mathcal{F}$ give a natural way to form (unconstrained) compatible pairs where F is also topologically tame, and hence the correspondence principle of Theorem 11.13 applies at every level $c \in \mathbb{R}$.

Theorem 11.20 (Canonical operators and compatible pairs). *Suppose that a subequation $\mathcal{F}$ admits a monotonicity cone subequation $\mathcal{M}$. Let $F \in C(\mathcal{J}^2)$ be the canonical operator for $\mathcal{F}$ determined by any fixed $J_0 \in \text{Int } \mathcal{M}$. Then*

(a) $(F, \mathcal{J}^2)$ *is a compatible proper elliptic operator–subequation pair;*
(b) $F(\mathcal{J}^2) = \mathbb{R}$ *and the operator F is topologically tame;*
(c) *for each $c \in \mathbb{R}$, the set $\mathcal{F}_c := \{ J \in \mathcal{J}^2 : F(J) \geq c \}$ is a subequation constraint set with $\mathcal{F}_0 = \mathcal{F}$ and the pair $(F, \mathcal{F}_c)$ satisfies the compatibility conditions*

$$\inf_{\mathcal{F}_c} F = c \quad \text{and} \quad \partial \mathcal{F}_c = \{ J \in \mathcal{F}_c : F(J) = c \}.$$

In addition, the canonical operator (determined by $J_0 \in \operatorname{Int} \mathcal{M}$) for the dual sub-equation $\widetilde{\mathcal{F}}$ is given by

$$\widetilde{F}(J) := -F(-J) \quad \text{for all } J \in \mathcal{J}^2, \text{ where also } \widetilde{\mathcal{F}}_c = \mathcal{F}_{-c}, \tag{11.46}$$

and statements analogous to (a), (b), and (c) hold for $(\widetilde{F}, \mathcal{J}^2)$ and $(\widetilde{F}, \widetilde{\mathcal{F}}_c)$.

Proof. The proper ellipticity of F on $\mathcal{J}^2$ is a consequence of Proposition 11.17(c). By Definition 11.1, the pair $(F, \mathcal{J}^2)$ is compatible if $F \in C(\mathcal{J}^2)$ and this follows from the fact that F is actually Lipschitz continuous, as will be proven in the following section. Next, as noted in the proof of Proposition 11.17(c), F is topologically tame in the sense of Definition 11.9 since the canonical operator satisfies the affine property (11.37). The affine property also shows that $F(\mathcal{J}^2) = \mathbb{R}$.

For part (c), each $\mathcal{F}_c$ is closed by the continuity of F and $\mathcal{F}_c$ is nonempty and not all of $\mathcal{J}^2$ since $F(\mathcal{J}^2) = \mathbb{R}$. Since the pair $(F, \mathcal{J}^2)$ is $\mathcal{M}$-monotone, each $\mathcal{F}_c$ is $\mathcal{M}$-monotone by Lemma 11.3 for the monotonicity cone subequation $\mathcal{M}$. Hence, $\mathcal{F}_c$ satisfies properties (P), (N), and (T) (see Proposition 4.7). The compatibility claim of part (c) follows from Theorem 11.10(4) and the continuity of F.

Finally, if $\mathcal{F}$ is $\mathcal{M}$-monotone then so is $\widetilde{\mathcal{F}}$ (using (4.12)) and since

$$-F(-J) := \min\{t \in \mathbb{R} : -J + tJ_0 \in \mathcal{F}\} := t_J$$

if $t < t_J$ one has

$$-J + tJ_0 \in \sim \operatorname{Int} \mathcal{F}, \quad \text{and hence } J - tJ_0 \in \widetilde{\mathcal{F}},$$

which shows that $-F(-J) = -\min\{t \in \mathbb{R} : J + tJ_0 \in \widetilde{\mathcal{F}}\}$, as needed. The remaining claims for $\widetilde{\mathcal{F}}_c$, $(\widetilde{F}, \mathcal{J}^2)$, and $(\widetilde{F}, \widetilde{\mathcal{F}}_c)$ then follow. $\qquad\square$

Combining the compatibility result of Theorem 11.20 with the correspondence principle and the general comparison result for subequations gives the following comparison principle for canonical operators.

Theorem 11.21 (Comparison for canonical operators). *Let $\mathcal{F} \subset \mathcal{J}^2$ be a subequation constraint set with monotonicity cone subequation $\mathcal{M}$. Suppose that $\mathcal{M}$ admits a strict approximator ψ on a bounded domain Ω, that is, $\psi \in C(\overline{\Omega}) \cap C^2(\Omega)$ such that $J_x^2 \psi \in \operatorname{Int} \mathcal{M}$ for each $x \in \Omega$. Then, for each $J_0 \in \operatorname{Int} \mathcal{M}$ fixed, the canonical operator F for $\mathcal{F}$ determined by J_0 satisfies the comparison principle at every level $c \in \mathbb{R}$, that is,*

$$u \leq w \text{ on } \partial\Omega \implies u \leq w \text{ on } \Omega \tag{11.47}$$

for $u \in \operatorname{USC}(\overline{\Omega})$ and $w \in \operatorname{LSC}(\overline{\Omega})$, which are respectively viscosity subsolutions and supersolutions to $F(u, Du, D^2u) = c$ on Ω.

Proof. By Theorem 11.20, we have that $(F, \mathcal{J}^2)$ is a compatible proper elliptic operator–subequation pair, every $c \in \mathbb{R}$ is an admissible level of F, and
F is topologically tame. Hence, by the correspondence principle of Theorem
11.13, at every level $c \in \mathbb{R}$ the comparison principle (11.47) holds for viscosity
subsolutions and supersolutions of the equation $F = c$ if and only if it holds for
$\mathcal{F}_c$-subharmonics and $\mathcal{F}_c$-superharmonics. Since $\mathcal{M}$ is a monotonicity cone for
each subequation $\mathcal{F}_c$ and since $\mathcal{M}$ admits a strict approximator ψ on Ω, comparison for $\mathcal{F}_c$ follows from the general comparison principle of Theorem 7.5.   □

We conclude this section with a few relevant examples and constructions.

Example 11.22. The pure second-order convexity subequation $\mathcal{F} = \mathbb{R} \times \mathbb{R}^n \times$
$\mathcal{P} = \mathcal{M}(\mathcal{P})$ is $\mathcal{M}$-monotone for $\mathcal{M} = \mathcal{M}(\mathcal{P})$ and the operator

$$F(J) = F(r, p, A) = \lambda_{\min}(A) \tag{11.48}$$

is easily seen to be the canonical operator for $\mathcal{F}$ determined by $J_0 = (0, 0, I)$ (or
any $J_0 = (r_0, p_0, I)$). Similarly,

$$\widetilde{F}(J) = \widetilde{F}(r, p, A) = \lambda_{\max}(A) \tag{11.49}$$

is the canonical operator (with the same J_0) for the subaffine subequation $\widetilde{\mathcal{F}} = \mathbb{R} \times \mathbb{R}^n \times \widetilde{\mathcal{P}}$, which is also $\mathcal{M}(\mathcal{P})$-monotone.

Families of subequations which are $\mathcal{M}$-monotone for a fixed convex cone
subequation $\mathcal{M}$ have particularly nice properties. For example, closed sets $\mathcal{F}$
which are $\mathcal{M}$-monotone automatically satisfy the topological property and are
subequations by Proposition 4.7 (see the discussion after the proposition for
counterexamples where monotonicity is lacking). Moreover, given an arbitrary
family of such subequations, by considering the dual family as well, intersections and unions lead to four associated subequations with computable canonical
operators, as follows.

Theorem 11.23 (Canonical operators, duality, intersections, and unions). *Sup-
pose that $\{\mathcal{F}_\sigma : \sigma \in \Sigma\}$ is an arbitrary family of subequations which are all $\mathcal{M}$-
monotone for a given monotonicity subequation cone $\mathcal{M}$. Let $J_0 \in \mathrm{Int}\,\mathcal{M}$ be fixed,
but arbitrary. Let $F_\sigma \in C(\mathcal{J}^2)$ denote the canonical operator (determined by J_0)
associated to the subequation $\mathcal{F}_\sigma$ and consider the dual operator $\widetilde{F}_\sigma \in C(\mathcal{J}^2)$
defined by $\widetilde{F}_\sigma(J) := -F_\sigma(-J)$, which is the canonical operator (determined by
J_0) for the dual ($\mathcal{M}$-monotone) subequation $\widetilde{\mathcal{F}}_\sigma$ by (11.46).*

(a) *The intersection $\mathcal{F} := \bigcap_{\sigma \in \Sigma} \mathcal{F}_\sigma$ (if nonempty) is an $\mathcal{M}$-monotone subequa-
    tion with canonical operator $F \in C(\mathcal{J}^2)$ (determined by J_0) given by the
    infimum*

$$F(J) := \inf_{\sigma \in \Sigma} F_\sigma(J), \quad J \in \mathcal{J}^2. \tag{11.50}$$

Applying this to the dual family $\{\widetilde{F}_\sigma\}_{\sigma \in \Sigma}$, *we have that*

(b) *the intersection* $\mathcal{G} := \bigcap_{\sigma \in \Sigma} \widetilde{F}_\sigma$ *(if nonempty) is an* $\mathcal{M}$-*monotone subequation with canonical operator* $G \in C(\mathcal{J}^2)$ *(determined by J_0) given by the infimum*

$$G(J) := \inf_{\sigma \in \Sigma} \widetilde{F}_\sigma(J), \quad J \in \mathcal{J}^2; \tag{11.51}$$

(c) *the closure of the union* $\overline{\bigcup_{\sigma \in \Sigma} \widetilde{F}_\sigma}$ *(if not equal to all of $\mathcal{J}^2$) is an* $\mathcal{M}$-*monotone subequation* $\mathcal{H}$ *with canonical operator* $H \in C(\mathcal{J}^2)$ *(determined by J_0) given by the supremum*

$$H(J) := \sup_{\sigma \in \Sigma} \widetilde{F}_\sigma(J), \quad J \in \mathcal{J}^2. \tag{11.52}$$

In fact,

(d) *the intersection* $\mathcal{G} := \bigcap_{\sigma \in \Sigma} \widetilde{F}_\sigma$ *and the closure of the union* $\overline{\bigcup_{\sigma \in \Sigma} \widetilde{F}_\sigma}$ *(with $\mathcal{G} \neq \emptyset$ and $\mathcal{H} := \widetilde{\mathcal{G}} \neq \mathcal{J}^2$) are dual subequations with dual canonical operators (determined by J_0) G and H.*

Applying this to $\mathcal{F} := \bigcap_{\sigma \in \Sigma} \mathcal{F}_\sigma$, *we have that*

(e) *the intersection* $\mathcal{F} := \bigcap_{\sigma \in \Sigma} \mathcal{F}_\sigma$ *and the closure of the union* $\overline{\bigcup_{\sigma \in \Sigma} \widetilde{F}_\sigma}$ *(with $\mathcal{F} \neq \emptyset$ and $\mathcal{E} := \widetilde{\mathcal{F}} \neq \mathcal{J}^2$) are dual subequations with dual canonical operators (determined by J_0) F and E, where*

$$E(J) := \sup_{\sigma \in \Sigma} \widetilde{F}_\sigma(J), \quad J \in \mathcal{J}^2. \tag{11.53}$$

Proof. We begin by noting that $\mathcal{F}$, $\mathcal{G}$, $\mathcal{H}$, and $\mathcal{E}$ are all $\mathcal{M}$-monotone subequations (if $\mathcal{F}$ and $\mathcal{G}$ are nonempty and $\mathcal{H}$ and $\mathcal{E}$ are not all of $\mathcal{J}^2$) by Proposition 4.8. Next we recall the following consequence of Propositions 11.17 and 11.18: given an $\mathcal{M}$-monotone subequation $\mathcal{F}$ and given $J_0 \in \text{Int}\,\mathcal{M}$, an operator F is the canonical operator for $\mathcal{F}$ (determined by J_0) if and only if one has both the affine property

$$F(J + t J_0) = F(J) + t \quad \text{for all } J \in \mathcal{J}^2, t \in \mathbb{R} \tag{11.54}$$

and the structural relations

$$\begin{array}{ll} \text{(a) } \partial \mathcal{F} = \{F(J) = 0\}, & \text{(b) } \text{Int}\,\mathcal{F} = \{F(J) > 0\}, \\ \text{(c) } \mathcal{J}^2 \setminus \mathcal{F} = \{F(J) < 0\}, & \end{array} \tag{11.55}$$

where it is sufficient to have (11.54) and only one of the conditions in (11.55).

For part (a), it remains to show that the inf operator F of (11.50) is the canonical operator of $\mathcal{F}$ (determined by $J_0 \in \operatorname{Int} \mathcal{M}$). Since each F_σ is canonical for $\mathcal{F}_\sigma$, we have (11.54) for F_σ, which then implies the validity of (11.54) for the infimum $F := \inf_{\sigma \in \Sigma} F_\sigma$. We will show that (11.55)(c) holds. We have

$$\sim\!\mathcal{F} = \sim\!\left(\bigcap_{\sigma\in\Sigma} \mathcal{F}_\sigma \right) = \{ J \in \mathcal{J}^2 : F_\sigma < 0 \text{ for some } \sigma \in \Sigma \}$$

$$= \{ J \in \mathcal{J}^2 : F(J) < 0 \}.$$

This completes the proof of part (a), and also of part (b) for the dual family.

The proofs of parts (c) and (d) are intertwined. Start from part (b) with the subequation $\mathcal{G} := \bigcap_{\sigma\in\Sigma} \widetilde{\mathcal{F}}_\sigma$ and canonical operator $G := \inf_{\sigma\in\Sigma} \widetilde{F}_\sigma$ for $\mathcal{G}$. Next we prove that the canonical operator $\widetilde{G}$ for the dual subequation $\widetilde{\mathcal{G}}$ is the operator H defined by (11.52). That is, $\widetilde{G} = H$. Using the definitions, we have

$$\widetilde{G}(J) := -G(-J) := -\inf_{\sigma\in\Sigma} \widetilde{F}_\sigma(-J)$$

$$:= -\inf_{\sigma\in\Sigma} (-F_\sigma(J))$$

$$= \sup_{\sigma\in\Sigma} F_\sigma(J) := H(J) \quad \text{for all } J \in \mathcal{J}^2. \tag{11.56}$$

Since $H = \widetilde{G}$ is the canonical operator for the dual subequation $\mathcal{H} := \widetilde{\mathcal{G}}$, to complete the proofs of (c) and (d) we must show that

$$\mathcal{H} = \overline{\bigcup_{\sigma\in\Sigma} \mathcal{F}_\sigma}. \tag{11.57}$$

Since H is the canonical operator for the $\mathcal{M}$-monotone subequation $\mathcal{H}$, conditions (a)–(c) of (11.55) applied to H yield

$$\mathcal{H} = \{ J \in \mathcal{J}^2 : H(J) := \sup_{\sigma\in\Sigma} F_\sigma(J) \geq 0 \}. \tag{11.58}$$

Hence, $\mathcal{F}_\sigma := \{ J \in \mathcal{J}^2 : F_\sigma(J) \geq 0 \} \subset \mathcal{H}$ for each $\sigma \in \Sigma$, which shows that $\overline{\bigcup_{\sigma\in\Sigma} \mathcal{F}_\sigma} \subset \mathcal{H}$, since $\mathcal{H} = \widetilde{\mathcal{G}}$ is closed. For the containment $\mathcal{H} \subset \overline{\bigcup_{\sigma\in\Sigma} \mathcal{F}_\sigma}$, it suffices to show that $\operatorname{Int} \mathcal{H} \subset \overline{\bigcup_{\sigma\in\Sigma} \mathcal{F}_\sigma}$ since the subequation $\mathcal{H}$ satisfies the topological property $\mathcal{H} = \overline{\operatorname{Int} \mathcal{H}}$. Now, by (11.55)(b) applied to H, we have

$$J \in \operatorname{Int} \mathcal{H} \iff H(J) := \sup_{\sigma\in\Sigma} F_\sigma(J) > 0 \iff F_{\sigma'}(J) > 0 \text{ for some } \sigma' \in \Sigma,$$

in which case $J \in \operatorname{Int} \mathcal{F}_{\sigma'}$, by (11.55)(b) applied to $F_{\sigma'}$. In fact, this proves that $\operatorname{Int} \mathcal{H} \subset \bigcup_{\sigma\in\Sigma} \operatorname{Int} \mathcal{F}_\sigma$ so that $\operatorname{Int} \mathcal{H} = \bigcup_{\sigma\in\Sigma} \operatorname{Int} \mathcal{F}_\sigma$. This completes parts (c) and (d).

Finally, part (d) applied to $\mathcal{F} := \bigcap_{\sigma \in \Sigma} \mathcal{F}_\sigma$ immediately gives part (e).   □

Note that when Σ is a finite index set, the inf in (11.50) becomes a minimum. Interesting examples come from the gradient-free case, where we note that the zeroth-order negativity subequation $\mathcal{M}(\mathcal{N}) = \mathcal{N} \times \mathbb{R}^n \times \mathcal{S}(n)$ has canonical operator (with $J_0 + (-1, p_0, A_0)$) given by

$$F(r, p, A) = -r. \tag{11.59}$$

Example 11.24. The gradient-free negative-convex subequation $\mathcal{F} = \mathcal{N} \times \mathbb{R}^n \times \mathcal{P} = \mathcal{M}(\mathcal{N}) \cap \mathcal{M}(\mathcal{P})$ is $\mathcal{M}$-monotone for $\mathcal{M} = \mathcal{M}(\mathcal{N}) \cap \mathcal{M}(\mathcal{P})$ and the operator

$$F(J) = F(r, p, A) = \min\{-r, \lambda_{\min}(A)\} \tag{11.60}$$

is easily seen to be the canonical operator for $\mathcal{F}$ determined by $J_0 = (-1, 0, I)$ (or any $J_0 = (-1, p_0, I)$). By duality (11.46), one can compute the canonical operator for the dual subequation to find

$$\widetilde{F}(r, p, A) = \max\{-r, \lambda_{\max}(A)\}$$

as the canonical operator for the subaffine-plus subequation

$$\widetilde{\mathcal{F}} = \{(r, p, A) \in \mathcal{J}^2 : r \leq 0 \text{ or } A \in \widetilde{\mathcal{P}}\}.$$

11.5 LIPSCHITZ REGULARITY OF SUBEQUATION BOUNDARIES

Monotonicity $\mathcal{M}$ for a subequation $\mathcal{F}$ forces the associated "equation" $\partial \mathcal{F}$ to have Lipschitz regularity. The relationship (11.44) shows that the canonical operator F for $\mathcal{F}$ (determined by J_0) will be Lipschitz continuous on $\mathcal{J}^2$ if and only if one the graphing functions g is Lipschitz. These functions actually are 1-Lipschitz with respect to natural seminorms built from the graphing function of the monotonicity cone subequation $\mathcal{M}$ over W_0, if W_0 is chosen carefully. The geometric reason for this regularity is that $\mathcal{M}$-monotonicity for $\mathcal{F}$ means that the translates of the cones $\mathcal{M}$ and $-c\mathcal{M}$ with vertices on $\partial \mathcal{F}$ must lie in the epigraph of F and its complement, respectively. This cone pinching is the Lipschitz property.

To start, denote by $\|\cdot\|^+ \colon W_0 \to \mathbb{R}$ the graphing function for $\partial \mathcal{M}$ over W_0 given by Proposition 11.19. That is,

$$\partial \mathcal{M} = \{J' + \|J'\|^+ J_0 : J' \in W_0\}. \tag{11.61}$$

Using (11.43), one also has that $\mathcal{M}$ is the epigraph of $\|\cdot\|^+$, that is,

$$\mathcal{M} = \left\{ J' + t J_0 : \text{where } t \geq \|J'\|^+ \text{ and } J' \in W_0 \right\}. \tag{11.62}$$

Note that, since $\mathcal{M}$ is a cone, the function $\|\cdot\|^+$ is *positively homogeneous of degree 1*, that is,

$$\|tJ\|^+ = t\|J\|^+ \quad \text{for each } t \geq 0 \text{ and for each } J' \in W_0. \tag{11.63}$$

Since $\mathcal{M}$ is a convex cone, the function $\|\cdot\|^+$ is also *subadditive*, that is,

$$\|J + J'\|^+ \leq \|J\|^+ + \|J'\|^+ \quad \text{for each pair } J, J' \in W_0. \tag{11.64}$$

Such *sublinear functions* always come in pairs. More precisely, given $\|\cdot\|^+$ satisfying (11.63) and (11.64), by defining

$$\|J\|^- := \|-J\|^+ \quad \text{for each } J \in W_0, \tag{11.65}$$

the function $\|\cdot\|^-$ also satisfies (11.63) and (11.64). For certain choices of the transverse hyperplane W_0 to J_0, the functional $\|\cdot\|^+$ (as well as $\|\cdot\|^-$) is a *seminorm* on W_0; that is, in addition to the sublinearity (11.63) and (11.64) for $\|\cdot\|^+$, one also has

$$\|J\|^+ \geq 0 \quad \text{for all } J \in W_0. \tag{11.66}$$

This will be proven in Proposition 11.29 below, but first we show that $\mathcal{M}$-monotonicity of a subequation $\mathcal{F}$ is equivalent to a weak 1-Lipschitz condition on the graphing function g for $\partial\mathcal{F}$.

Proposition 11.25 (Lipschitz regularity of $\partial\mathcal{F}$). *Suppose that $\mathcal{F}$ is a subequation and that $\mathcal{M}$ is a convex cone subequation. Then $\mathcal{F} + \mathcal{M} \subset \mathcal{F}$ if and only if the graphing function g for $\partial\mathcal{F}$ (defined in Proposition 11.19) satisfies*

$$-\|J'\|^- \leq g(J + J') - g(J) \leq \|J'\|^+ \quad \text{for each pair } J, J' \in W_0. \tag{11.67}$$

Proof. Given $J, J' \in W_0$, by (11.42) and (11.61) we have

$$J + g(J)J_0 \in \partial\mathcal{F} \quad \text{and} \quad J' + \|J'\|^+ J_0 \in \partial\mathcal{M}.$$

Assume that $\mathcal{F} + \mathcal{M} \subset \mathcal{F}$. Then the sum

$$J + J' + (g(J) + \|J'\|^+)J_0$$

belongs to $\mathcal{F}$. By (11.43) one has

$$g(J) + \|J'\|^+ \geq g(J + J'),$$

which is the right-hand inequality in the Lipschitz bound (11.67). This right-hand inequality implies the left-hand inequality. Replacing J' by $-J'$, the right-hand inequality in (11.67) can be restated as

$$g(J - J') - g(J) \leq \|-J'\|^+,$$

that is,

$$-\|J'\|^- \leq g(J) - g(J - J'),$$

which by relabeling is the left-hand inequality.

For the converse, assume that the Lipschitz bound (11.67) holds. By (11.43), the elements of $\mathcal{F}$ are all of the form $J + tJ_0$ with $J \in W_0$ and $t \geq g(J)$ and by (11.62) the elements of $\mathcal{M}$ are all of the form $J' + sJ_0$ with $J' \in W_0$ and $s \geq \|J'\|^+$. Therefore, the elements of $\mathcal{F} + \mathcal{M}$ are all of the form

$$J + J' + (t + s)J_0 \quad \text{with } J, J' \in W_0,\ t \geq g(J),\ \text{and } s \geq \|J'\|^+. \tag{11.68}$$

This jet belongs to $\mathcal{F}$ (again by (11.43)) if and only if

$$t + s \geq g(J + J'), \tag{11.69}$$

but by the second inequality in the Lipschitz estimate (11.67) we have

$$t + s \geq g(J) + \|J'\|^+ \geq g(J + J'),$$

which gives (11.69), as needed to conclude $\mathcal{F} + \mathcal{M} \subset \mathcal{F}$.   □

Assuming for the moment that $\|\cdot\|^+$ is a seminorm (by choosing a suitable hyperplane W_0), if one considers the seminorm $\|\cdot\|$ on W_0 defined by the sum $\|\cdot\| := \|\cdot\|^+ + \|\cdot\|^-$, then estimate (11.67) yields the Lipschitz estimate

$$|g(J + J') - g(J)| \leq \|J'\| \quad \text{for each pair } J, J' \in W_0, \tag{11.70}$$

which completes the claim that g and hence F are continuous for Theorem 11.20.

Finally, we show how to choose W_0 so that $\|\cdot\|^+ \geq 0$ on W_0, which is the remaining seminorm property (11.66). This is a general fact about finite-dimensional inner product spaces $(V, \langle\cdot, \cdot\rangle)$ of which the 2-jet space $\mathcal{J}^2$ is an example. Consider a closed convex cone $\mathcal{M} \subset V$ (with vertex at the origin). The *edge* E of $\mathcal{M}$ is defined to be

$$E := \mathcal{M} \cap (-\mathcal{M}), \tag{11.71}$$

and one can show that the vector subspace E contains all other vector subspaces of $\mathcal{M}$ (where $E = \{0\}$ is possible). The (*convex cone*) *polar* $\mathcal{M}^\circ$ of $\mathcal{M}$ is defined by

$$\mathcal{M}^\circ := \{J \in V : \langle J, J' \rangle \geq 0 \text{ for each } J' \in \mathcal{M}\}. \tag{11.72}$$

Recall that the *bipolar theorem* says that the polar of $\mathcal{M}^\circ$ is $\mathcal{M}$. Let S denote the *span of* $\mathcal{M}^\circ$ in V. If W is any linear subspace, its polar $W^\circ = W^\perp$ is just its orthogonal complement. This notion of polar will also be used in Section 12.4 when we discuss Hamilton–Jacobi–Bellman operators and we record a few observations now.

Remark 11.26. In a finite-dimensional inner product space $(V, \langle \cdot, \cdot \rangle)$, given any subset $T \subset V$ one can define its (convex cone) polar as in (11.72), that is,

$$T^\circ := \{J \in V : \langle J, J' \rangle \geq 0 \text{ for each } J' \in T\}. \tag{11.73}$$

One knows that T° is always a closed convex cone and that

$$T^\circ = C(T)^\circ, \quad \text{where } C(T) \text{ is the closed convex hull of } T. \tag{11.74}$$

Returning to the judicious choice of the needed hyperplane W_0, we will need the following fact.

Lemma 11.27. *With V, E, and S as above, one has that*

$$V = E \oplus S \text{ is an orthogonal decomposition.} \tag{11.75}$$

Proof. If $e \in E$ and $J \in \mathcal{M}_+$, then $\pm e \in \mathcal{M}$ and hence $\pm\langle e, J \rangle \geq 0$, that is, $\langle e, J \rangle = 0$ and hence $E \perp S$. Now $\mathcal{M}_+ \subset S$ implies that the polars satisfy $S^\perp \subset (\mathcal{M}_+)_+ = \mathcal{M}$. Hence, the linear subspace S satisfies $S^\perp \subset E$ so that $V = S^\perp \oplus S \subset E \oplus S$, which forces (11.75). $\square$

We are now ready to describe the judicious choice of the hyperplane W_0. Pick $J_0 \in \text{Int}\,\mathcal{M} \neq \emptyset$ as in the structure theorem. Now pick the hyperplane W_0 to have normal $J_0' \in \text{Int}_\text{rel}\,\mathcal{M}_+$, the interior of the convex cone $\mathcal{M}_+$ relative to the vector space $S = \text{span}\,\mathcal{M}_+$. Of course, since $J_0' \in \mathcal{M}_+$, the original convex cone $\mathcal{M}$ satisfies

$$\mathcal{M} \subset H, \tag{11.76}$$

where H is the closed half-space

$$H := \{J \in V : \langle J, J_0' \rangle \geq 0\}, \tag{11.77}$$

whose boundary satisfies

$$\partial H = W_0. \tag{11.78}$$

Lemma 11.28. *Suppose that W_0 is the hyperplane transversal to J_0 with normal $J_0' \in \mathrm{Int}_{\mathrm{rel}}\, \mathcal{M}_+$ as described above. Then one has*

$$\mathcal{M} \cap W_0 = E \tag{11.79}$$

and

$$\langle J_0', J_0 \rangle > 0, \tag{11.80}$$

which implies the transversality $W_0 \pitchfork J_0$.

Proof. For the relation (11.79), we begin by noting that the relation $E \subset \mathcal{M} \cap S^\perp \subset \mathcal{M} \cap W_0$ was established above. Now suppose that $J \in \mathcal{M} \cap W_0$ and using (11.75) decompose $J := J_E + J_S$ with $J_E \in E$ and $J_S \in S$. Since $J_0' \in \mathrm{Int}_{\mathrm{rel}}\, \mathcal{M}_+$ and $J_S \in S$, if $\varepsilon > 0$ is sufficiently small then $J_0' - \varepsilon J_S \in \mathcal{M}_+$. Finally, since $J \in \mathcal{M} \cap W_0$ one has

$$0 \leq \langle J, J_0' - \varepsilon J_S \rangle = \langle J, J_S \rangle - \varepsilon \langle J, J_S \rangle = -\varepsilon \langle J_S, J_S \rangle,$$

proving $J_S = 0$ and so $J = J_E \in E$ as desired.

For the transversality (11.80), with $J_0' \in \mathcal{M}_+$, $J_0 \in \mathcal{M}$, one has $\langle J_0', J_0 \rangle$, where $J_0 \in \mathrm{Int}\, \mathcal{M}$ easily implies that $\langle J_0', J_0 \rangle > 0$, as above. $\square$

The needed seminorm property (11.66) follows from the above considerations.

Proposition 11.29. *Let $\mathcal{M}$ be a convex cone subequation with $J_0 \in \mathrm{Int}\, \mathcal{M}$ fixed. Suppose that W_0 is the hyperplane transversal to J_0 with normal $J_0' \in \mathrm{Int}_{\mathrm{rel}}\, \mathcal{M}_+$ as in Lemma 11.28 and let $\| \cdot \|^+$ be the graphing function of $\partial \mathcal{M}$ over W_0. Then one has*

$$\|J\|^+ \geq 0 \quad \text{for each } J \in W_0 \tag{11.81}$$

and for each $J \in W_0$ we have

$$\|J\|^+ = 0 \iff J \in E = \mathcal{M} \cap (-\mathcal{M}). \tag{11.82}$$

Proof. It suffices to note that statements (11.81) and (11.82) are equivalent to the statements

$$\mathcal{M} \subset H := \{ J \in V : \langle J, J_0' \rangle \geq 0 \} \tag{11.83}$$

and

$$\mathcal{M} \cap \partial H = E \quad (\text{where } W_0 = \partial H), \tag{11.84}$$

which were noted in (11.76)–(11.78). $\square$

11.6   GÅRDING–DIRICHLET OPERATORS

In our dichotomy between constrained and unconstrained operator–subequation pairs, the constrained case is best illustrated by examples involving *Gårding–Dirichlet operators* $\mathfrak{g}$, and provides many interesting examples of compatible operator–subequation pairs illustrating the correspondence principle of Theorem 11.13. The most basic example is the classical Monge–Ampère operator $F(A) := \det A = \lambda_1(A) \cdots \lambda_n(A)$ on $\mathcal{S}(n)$. With the standard restriction of F to the convexity subequation $\mathcal{P} \subset \mathcal{S}(n)$, the pair $(\det, \mathcal{P})$ is a pure second-order compatible pair. This pair can be thought of as the "universal example" by applying the following procedure starting from any Gårding polynomial $\mathfrak{g}$ of degree m. Simply substituting the so-called *Gårding eigenvalues* $\lambda_k^{\mathfrak{g}}(A)$ for the standard eigenvalues yields the *Gårding–Dirichlet* operator $\mathfrak{g}(A) := \lambda_1^{\mathfrak{g}}(A) \cdots \lambda_m^{\mathfrak{g}}(A)$, and restricting $\mathfrak{g}$ to the *closed Gårding cone* $\overline{\Gamma}_{\mathfrak{g}}$ (the pull-back of $[0, +\infty)^m$ under the eigenvalue map $\lambda^{\mathfrak{g}}(A) := (\lambda_1^{\mathfrak{g}}(A), \ldots, \lambda_m^{\mathfrak{g}}(A)))$ yields a multitude of interesting examples (see Examples 11.33 and 11.34 below). In addition, this provides a unified approach to studying many of the most important pure second-order subequations and we refer the reader to [48, 51] for a modern, self-contained, and detailed treatment. In this section we will focus mainly on pure second-order operators $F(r, p, A) := \mathfrak{g}(A)$ with $\mathfrak{g}$ a Gårding–Dirichlet polynomial (see Definition 11.31), but it is important to note they give important building blocks for operators $F = F(r, p, A)$ which contain some $\mathfrak{g}(A)$ as a factor. This will play a key role in Chapter 12 on comparison principles for operators F in constrained cases. Moreover, we present a new construction in Lemma 11.35 below which produces gradient-free compatible Gårding–Dirichlet pairs from pure second-order compatible Gårding–Dirichlet pairs.

In what follows, let $\mathfrak{g}$ be a homogeneous polynomial of degree m on $\mathcal{S}(n)$. Suppose that $\mathfrak{g}$ is *I-hyperbolic*, that is, $\mathfrak{g}(I) > 0$ and, for any given $A \in \mathcal{S}(n)$, the one-variable polynomial $t \mapsto \mathfrak{g}(tI + A)$ has exactly m real roots, $t_k^{\mathfrak{g}}(A)$ for $k = 1, \ldots, m$, whose negatives $\lambda_k^{\mathfrak{g}}(A) := -t_k^{\mathfrak{g}}(A)$ are called the *Gårding eigenvalues, or the I-eigenvalues, of A*. Up to permutation, we can order the Gårding eigenvalues

$$\lambda_1^{\mathfrak{g}}(A) \le \lambda_2^{\mathfrak{g}}(A) \le \cdots \le \lambda_m^{\mathfrak{g}}(A), \tag{11.85}$$

where we will often denote $\lambda_1^{\mathfrak{g}}(A)$ by $\lambda_{\min}^{\mathfrak{g}}(A)$ and $\lambda_m^{\mathfrak{g}}(A)$ by $\lambda_{\max}^{\mathfrak{g}}(A)$. If we normalize $\mathfrak{g}$ to have $\mathfrak{g}(I) = 1$ then it factors as

$$\mathfrak{g}(tI + A) = \prod_{k=1}^{m} (t + \lambda_k^{\mathfrak{g}}(A)), \tag{11.86}$$

so that

$$\mathfrak{g}(A) = \prod_{k=1}^{m} \lambda_k^{\mathfrak{g}}(A) \quad \text{and} \quad \lambda_k^{\mathfrak{g}}(A + sI) = \lambda_k^{\mathfrak{g}}(A) + s, \quad k = 1, \ldots, m. \tag{11.87}$$

Note that by the product formula (11.87), every Gårding–Dirichlet operator $\mathfrak{g}$ is a *generalized Monge–Ampère operator*, where the standard eigenvalues $\lambda_k(A)$ for the special case $\mathfrak{g} = \det$ are replaced by the Gårding I-eigenvalues $\lambda_k^{\mathfrak{g}}(A)$ for a general I-hyperbolic polynomial $\mathfrak{g}$.

The (*open*) *Gårding cone* Γ can be defined by

$$\Gamma := \left\{ A \in \mathcal{S}(n) : \lambda_{\min}^{\mathfrak{g}}(A) > 0 \right\}. \tag{11.88}$$

The conditions (11.87) easily imply that the closed Gårding cone satisfies

$$\overline{\Gamma} := \left\{ A \in \mathcal{S}(n) : \lambda_{\min}^{\mathfrak{g}}(A) \geq 0 \right\} \quad \text{and} \quad \partial\Gamma = \left\{ A \in \overline{\Gamma} : \mathfrak{g}(A) = 0 \right\}. \tag{11.89}$$

Gårding's theory includes two important results, namely the convexity of the Gårding cone Γ and the strict Γ-monotonicity of the Gårding eigenvalues.

Theorem 11.30 (Gårding, [40]). *Suppose that $\mathfrak{g}$ is an I-hyperbolic polynomial of degree m on $\mathcal{S}(n)$. Then the Gårding cone Γ is an open convex cone with vertex at the origin and the ordered I-eigenvalues of A are strictly Γ-monotone; that is, for each $k = 1, \ldots, m$,*

$$\lambda_k^{\mathfrak{g}}(A + B) > \lambda_k^{\mathfrak{g}}(A) \quad \text{for each } A \in \mathcal{S}(n), \ B \in \Gamma. \tag{11.90}$$

The statement of Theorem 11.30 combines [51, Theorem 5.1] on convexity (which is also shown to be equivalent to the convexity of $\lambda_{\max}^{\mathfrak{g}}(A) = -\lambda_{\min}^{\mathfrak{g}}(-A)$, or of the concavity of $\lambda_{\min}^{\mathfrak{g}}$) and [51, Theorem 6.2] on monotonicity. The reader is referred to [51] for the proofs.

The closed Gårding cone $\overline{\Gamma}$ is a closed convex cone with nonempty interior Γ, but it must also satisfy the positivity condition

$$\overline{\Gamma} + \mathcal{P} \subset \overline{\Gamma} \quad \text{(equivalently, } \mathcal{P} \subset \overline{\Gamma}) \tag{11.91}$$

in order to be a subequation. (Perhaps it is worth mentioning the fact that the requirement (11.91) implies that $\mathfrak{g}$ is hyperbolic in every positive definite direction $P > 0$ in $\mathcal{S}(n)$.)

Definition 11.31. A homogeneous polynomial $\mathfrak{g}$ on $\mathcal{S}(n)$ which is I-hyperbolic and for which the closed Gårding cone satisfies positivity $\overline{\Gamma} + \mathcal{P} \subset \overline{\Gamma}$ will be called a *Gårding–Dirichlet polynomial*.

Gårding–Dirichlet polynomials yield a rich class of important examples of pure second-order operators and subequations which are amenable to the theory developed in [46]. As noted above, they will be exploited in Chapter 12 to illustrate comparison principles for second-order operators F which have some $\mathfrak{g}(A)$ as a factor. In the pure second-order case, we record the following elementary properties.

Proposition 11.32. *Suppose that $\mathfrak{g}$ is a Gårding–Dirichlet polynomial on $\mathcal{S}(n)$ with closed Gårding cone $\overline{\Gamma}$. By restricting $\mathfrak{g}$ to $\overline{\Gamma}$, one has*

(a) $(\mathfrak{g}, \overline{\Gamma})$ *is a compatible constrained operator–subequation pair;*
(b) *the operator $\mathfrak{g}$ is tame on $\overline{\Gamma}$; in fact,*

$$\exists\, C > 0 \text{ such that } \quad \mathfrak{g}(A + tB) > \mathfrak{g}(A) + Ct^{1/m} \quad \forall\, A \in \overline{\Gamma},\ B \in \Gamma; \quad (11.92)$$

(c) *the operator $\mathfrak{g}$ is topologically tame on $\overline{\Gamma}$.*

Proof. The compatibility of part (a) is immediate from (11.89). For the proof of tameness (b), we refer the reader to [60, Proposition 6.11], which makes use of (11.87). Tame implies topologically tame. But also note that topological tameness is immediate since $\mathfrak{g}$ is real analytic.   □

Properties (a) and (c) in Proposition 11.32 imply that

$$\text{comparison for the operator } \mathfrak{g}|_{\overline{\Gamma}} \ \Leftrightarrow\ \text{comparison for the subequation } \overline{\Gamma},$$
$$(11.93)$$

by Theorem 11.13. Property (b) plays a key role in comparison on domains Ω for inhomogeneous equations $\mathfrak{g}(D^2 u) = \psi(x)$, with ψ a continuous function on Ω (see [60]).

Now we turn to listing some of these interesting and important examples.

Examples 11.33 (Pure second-order Gårding–Dirichlet operator–subequation pairs). Let $\lambda_1(A) \leq \cdots \leq \lambda_n(A)$ denote the ordered eigenvalues of $A \in \mathcal{S}(n)$. As noted above, the most basic example is the following:

(1) (The elementary Monge–Ampère pair.) $\mathfrak{g}(A) := \det A$ and $\overline{\Gamma} = \mathcal{P}$.

There are many others, including the following examples:

(2) (The k-Hessian pair). For $k = 1, 2, \ldots, n$, denote by

$$\sigma_k(\lambda) := \sum_{1 \leq i_1 < \cdots < i_k \leq n} (\lambda_{i_1} \cdots \lambda_{i_k}) \quad \text{for all } \lambda = (\lambda_1, \ldots, \lambda_n) \in \mathbb{R}^n,$$

the kth elementary symmetric function of $\lambda \in \mathbb{R}^n$. The k-Hessian pair is

$$\mathfrak{g}(A) := \sigma_k(\lambda(A)) \quad \text{and} \quad \overline{\Gamma} = \{\sigma_j(\lambda(A)) \geq 0,\ j = 1, \ldots, k\}.$$

Here the Gårding I-eigenvalues of $\mathfrak{g}$ have no explicit formula in terms of the standard eigenvalues for $1 < k < n$, but they are real, since the roots of $\sigma_k(\lambda(A + tI))$ are critical points of $\sigma_{k+1}(\lambda(A + tI))$. Of course, when $k = n$ one recovers the Monge–Ampère operator (1). The notion of a principal eigenvalue for this pair was recently studied in [10].

(3) (The geometric k-convexity pair). This example was introduced and studied in [46]. It is geometrically significant because the plurisubharmonics (or simply subharmonics or subsolutions) are precisely the upper-semicontinuous functions that restrict to all affine k-planes to be classically (Laplacian) subharmonic. For $k = 1, 2, \ldots, n$, consider the symmetric polynomial

$$\tau_k(\lambda) := \prod_{1 \leq i_1 < \cdots < i_k \leq n} (\lambda_{i_1} + \cdots + \lambda_{i_k}) \quad \text{for all } \lambda = (\lambda_1, \ldots, \lambda_n) \in \mathbb{R}^n.$$

The geometric k-convexity pair is

$$\mathfrak{g}(A) := \tau_k(\lambda(A)) \quad \text{and} \quad \overline{\Gamma} = \{\lambda_1(A) + \cdots + \lambda_k(A) \geq 0\}.$$

The Gårding I-eigenvalues of $\mathfrak{g}$ are the pull-backs to $\mathcal{S}(n)$ (under the eigenvalue map $\lambda \colon \mathcal{S}(n) \to \mathbb{R}$) of the factors in the above product. In particular, the minimum Gårding I-eigenvalue is $\lambda_1(A) + \cdots + \lambda_k(A)$, which is the canonical operator for $\overline{\Gamma}$. This canonical operator was recently studied in [7], where the name *truncated Laplacian* was introduced.

(4) (The Lagrangian plurisubharmonic Monge–Ampère pair). This is a new pair, introduced in [46] and the main object of study in [58]. Its subharmonics are those upper-semicontinuous functions whose restrictions to arbitrary Lagrangian n-planes in $\mathbb{C}^n$ are classically (Laplacian) subharmonic. The closed Gårding cone can be defined by

$$\overline{\Gamma} := \{A \in \mathcal{S}(2n) : \operatorname{tr}(A|_L) \geq 0 \text{ for all Lagrangian } n\text{-planes } L \text{ in } \mathbb{C}^n\}.$$

However, a description of the Gårding–Dirichlet operator $\mathfrak{g}$ is somewhat involved. We encourage the reader to consult [58] for details, including the Lagrangian pluripotential theory.

Also, there are versions of (1), (2), and (3) over $\mathbb{C}$ or $\mathbb{H}$ instead of $\mathbb{R}$. See [48, 51] for a detailed discussion.

Examples 11.34 (Constructing more examples). We describe three standard methods for constructing Gårding–Dirichlet polynomials from a given Gårding–Dirichlet polynomial $\mathfrak{g}$. Suppose that $\mathfrak{g}$ is I-hyperbolic of degree m with ordered Gårding I-eigenvalues $\lambda_1^{\mathfrak{g}}(A) \leq \cdots \leq \lambda_m^{\mathfrak{g}}(A)$ and Gårding cone $\Gamma_{\mathfrak{g}}$. The first two methods generalize the constructions given in examples (2) and (3) above.

(I) (Partial derivatives in the direction I/elementary symmetric functions). For each $k = 0, 1, \ldots, m$, the degree $m - k$ polynomial

$$\mathfrak{g}_k(A) := \frac{d^k}{dt^k} \mathfrak{g}(A + tI)|_{t=0} = \sigma_{m-k}(\lambda^{\mathfrak{g}}(A)) \quad \text{(modulo a positive rescaling)}$$

is also a Gårding–Dirichlet polynomial which is I-hyperbolic, whose open Gårding cones are nested:

$$\Gamma_{\mathfrak{g}} \subset \cdots \subset \Gamma_{\mathfrak{g}_k} \quad \text{with } \mathcal{P} \subset \overline{\Gamma}_{\mathfrak{g}_k}. \tag{11.94}$$

When $\mathfrak{g} := \det$, this procedure produces (2) above.

(II) (k-fold sums of Gårding eigenvalues). For each $k = 1, \ldots, m$, the degree $\binom{m}{k}$ polynomial

$$\mathfrak{g}_k(A) := \prod_{1 \leq i_1 < \cdots < i_k \leq n} \left(\lambda_{i_1}^{\mathfrak{g}}(A) + \cdots + \lambda_{i_k}^{\mathfrak{g}}(A) \right)$$

is also a Gårding–Dirichlet polynomial which is I-hyperbolic with Gårding cones $\Gamma_{\mathfrak{g}_k}$ that satisfy (11.94). The Gårding I-eigenvalues of $\mathfrak{g}_k$ are the factors in the above product. When $\mathfrak{g} := \det$, one has example (3) above.

Note that method (I) decreases the degree m, while method (II) increases the degree m. There is a third method, which has its origins in the work of Krylov (see [77, Definition 2.13]) and leaves the degree of $\mathfrak{g}$ fixed.

(III) (ε-(uniformly) elliptic regularization). For $\varepsilon > 0$, the degree-m polynomial

$$\mathfrak{g}_\varepsilon := \prod_{k=1}^{m} (\lambda_k^{\mathfrak{g}}(A) + \varepsilon \operatorname{tr}^{\mathfrak{g}}(A)),$$

where $\operatorname{tr}^{\mathfrak{g}}(A) := \sum_{k=1}^{m} \lambda_k^{\mathfrak{g}}(A)$, is a Gårding–Dirichlet polynomial.

The reader is referred to [48, Section 5] for additional details on these three methods.

Next we introduce a new construction, which produces a Gårding–Dirichlet polynomial $\mathfrak{h}$ of degree m on $\mathbb{R} \times \mathcal{S}(n)$ from a Gårding–Dirichlet polynomial $\mathfrak{g}$ of degree m on $\mathcal{S}(n)$ by cleverly "adding a real variable $r \in \mathbb{R}$." This leads to new gradient-free compatible proper elliptic pairs $(\mathfrak{h}, \overline{\Gamma}_{\mathfrak{h}})$.

Lemma 11.35. *Suppose that $\mathfrak{g}$ is a Gårding–Dirichlet polynomial on $\mathcal{S}(n)$ which is hyperbolic in the direction I of degree m and with Gårding eigenvalues $\lambda_k^{\mathfrak{g}}(A)$ ($k = 1, \ldots, m$) and Gårding cone $\Gamma_{\mathfrak{g}}$ (normalized to have $\mathfrak{g}(I) = 1$). Define the degree-m polynomial $\mathfrak{h}$ on $\mathbb{R} \times \mathcal{S}(n)$ by*

$$\mathfrak{h}(r, A) := \mathfrak{g}(A - rI), \quad (r, A) \in \mathbb{R} \times \mathcal{S}(n). \tag{11.95}$$

Then $\mathfrak{h}$ is $(-\frac{1}{2}, \frac{1}{2}I)$-hyperbolic with Gårding eigenvalues

$$\lambda_k^{\mathfrak{h}}(r, A) := \lambda_k^{\mathfrak{g}}(A) - r, \quad k = 1, \ldots, m, \tag{11.96}$$

and Gårding cone

$$\Gamma_\mathfrak{h} := \big\{(r, A) \in \mathbb{R} \times \mathcal{S}(n) : A - rI \in \Gamma_\mathfrak{g}\big\} = \big\{(r, A) : r \leq \lambda_1^\mathfrak{g}(A)\big\}. \qquad (11.97)$$

Moreover,

$$\overline{\Gamma}_\mathfrak{g} \text{ is } \mathcal{P}\text{-monotone} \ \Leftrightarrow \ \overline{\Gamma}_\mathfrak{h} \text{ is } (\mathcal{N} \times \mathcal{P})\text{-monotone}, \qquad (11.98)$$

and hence $(\mathfrak{h}, \overline{\Gamma}_\mathfrak{h})$ *is a compatible gradient-free subequation pair.*

Proof. First, notice that for each $t \in \mathbb{R}$ and each $(r, A) \in \mathbb{R} \times \mathcal{S}(n)$ one can easily show that

$$\mathfrak{h}\big(t(-\tfrac{1}{2}, \tfrac{1}{2}I) + (r, A)\big) = \prod_{k=1}^m (t + \lambda_k^\mathfrak{g}(A - rI)) = \prod_{k=1}^m (t + (\lambda_k^\mathfrak{g}(A) - r)),$$

so that $\mathfrak{h}$ is $(-\tfrac{1}{2}, \tfrac{1}{2}I)$-hyperbolic with Gårding eigenvalues as claimed in (11.96) and Gårding cone as claimed in (11.97).

Now, since the closed Gårding cones are convex,

$$\overline{\Gamma}_\mathfrak{g} \text{ is } \mathcal{P}\text{-monotone} \ \Leftrightarrow \ \mathcal{P} \subset \overline{\Gamma}_\mathfrak{g}$$

and

$$\overline{\Gamma}_\mathfrak{h} \text{ is } (\mathcal{N} \times \mathcal{P})\text{-monotone} \ \Leftrightarrow \ \mathcal{N} \times \mathcal{P} \subset \overline{\Gamma}_\mathfrak{h}.$$

To complete the proof of (11.98), note that

$$\lambda_1^\mathfrak{g}(A) - r \geq 0 \ \forall\, (r, A) \in \mathcal{N} \times \mathcal{P} \ \Leftrightarrow \ \lambda_1^\mathfrak{g}(A) \geq 0 \ \forall\, A \in \mathcal{P}. \qquad \square$$

It is also of interest that, just as in the pure second-order case,

$$\mathcal{H} := \big\{(r, A) \in \mathbb{R} \times \mathcal{S}(n) : \mathfrak{h}(r, A) = 0\big\}$$

has m *subequation branches*

$$\big\{(r, A) \in \mathbb{R} \times \mathcal{S}(n) : r \leq \lambda_k^\mathfrak{g}(A)r\big\}, \quad k = 1, \ldots, m,$$

with principal (smallest) branch

$$\Lambda_1^\mathfrak{g} := \big\{(r, A) \in \mathbb{R} \times \mathcal{S}(n) : r \leq \lambda_1^\mathfrak{g}(A)\big\} = \overline{\Gamma}.$$

This important notion of branches will be further developed in the next section. To get started, consider the pure second-order equation

$$\mathcal{H} := \big\{A \in \mathcal{S}(n) : \mathfrak{g}(A) = 0\big\} \qquad (11.99)$$

and its *principal branch* (see (11.89))

$$\Lambda_1^{\mathfrak{g}} := \left\{ A \in \mathcal{S}(n) : \lambda_1^{\mathfrak{g}}(A) \geq 0 \right\} = \overline{\Gamma}. \tag{11.100}$$

Since $\Lambda_1^{\mathfrak{g}}$ is a convex cone, it is automatically $\overline{\Gamma} = \Lambda_1^{\mathfrak{g}}$-monotone. Now, $\mathcal{P} \subset \overline{\Gamma}$ implies that this principal branch $\Lambda_1^{\mathfrak{g}}$ is a pure second-order subequation which is topologically tame, and hence comparison holds on arbitrary bounded domains. Since $\mathcal{P} \subset \overline{\Gamma}$, by the monotonicity in Gårding's theorem (Theorem 11.30), each of the other branches

$$\Lambda_k^{\mathfrak{g}} := \left\{ A \in \mathcal{S}(n) : \lambda_k^{\mathfrak{g}}(A) \geq 0 \right\}, \quad k = 2, \dots, m \tag{11.101}$$

are subequations. Note that $\Lambda_1^{\mathfrak{g}} \subset \Lambda_2^{\mathfrak{g}} \subset \cdots \subset \Lambda_m^{\mathfrak{g}}$, and since $-\lambda_k^{\mathfrak{g}}(-A) = \lambda_{m-k+1}^{\mathfrak{g}}(A)$ the dual subequation is $\widetilde{\Lambda}_k^{\mathfrak{g}} = \Lambda_{m-k+1}$. The *branch condition* means that $\partial \Lambda_k^{\mathfrak{g}} \subset \mathcal{H}$. The second part of (11.87) says that $\lambda_k^{\mathfrak{g}}$ is the canonical operator for $\Lambda_k^{\mathfrak{g}}$.

The most basic example is $\mathfrak{g}(A) = \det(A)$, in which case $\lambda_k^{\mathfrak{g}}(A) = \lambda_k(A)$ for $k = 1, \dots, n$ and $\overline{\Gamma} = \mathcal{P}$.

11.7   SUBEQUATION BRANCHES

We conclude this chapter with the notion of subequation branches. This notion makes sense for any closed set $\mathcal{H}$ contained in the jet space $\mathcal{J}^2$ and is independent of the existence of an operator $F \in C(\mathcal{F}, \mathbb{R})$ whose zero level set satisfies $\{F = 0\} = \mathcal{H}$, that is, the case where $\mathcal{H}$ can be thought of as an *equation constraint set* for the partial differential equation $F(u, Du, D^2 u) = 0$.

Definition 11.36. Suppose that $\mathcal{H} \subset \mathcal{J}^2$ is any closed set. A subequation $\mathcal{F}_0 \subset \mathcal{J}^2$ is called a *subequation branch* of $\mathcal{H}$ if $\partial \mathcal{F}_0 \subset \mathcal{H}$.

In general, an equation constraint set $\mathcal{H}$ may admit more than one subequation branch. We have seen several examples in the previous section. A standard pure first-order example is the Eikonel equation $\mathcal{H} = \{p \in \mathbb{R}^n : |p| = 1\}$. It admits two subequation branches

$$\mathcal{F}^- = \left\{ p \in \mathbb{R}^n : |p| \leq 1 \right\} \quad \text{and} \quad \mathcal{F}^+ = \left\{ p \in \mathbb{R}^n : |p| \geq 1 \right\},$$

whose boundaries are $\mathcal{H}$. However, if one subequation branch $\mathcal{F}$ of $\mathcal{H}$ has monotonicity $\mathcal{M}$ which is a subequation and if $\partial \mathcal{F}$ is all of $\mathcal{H}$, then the equation $\mathcal{H}$ uniquely determines the subequation branch $\mathcal{F}$ as another consequence of the structure theorem (Theorem 11.14). The precise statement is as follows.

Proposition 11.37. *If a subequation $\mathcal{F} \subsetneq \mathcal{J}^2$ admits a monotonicity cone subequation $\mathcal{M}$, then one knows by Corollary 11.15 that $\mathcal{F} = \partial \mathcal{F} + \mathcal{M}$. Consequently, if $\mathcal{H} = \partial \mathcal{F}$ and $\mathcal{H} = \partial \mathcal{F}'$, where $\mathcal{F}'$ is also an $\mathcal{M}$-monotone subequation, then $\mathcal{F} = \mathcal{F}'$.*

Part IV

Applications to PDEs

Chapter Twelve

Comparison Principles for Operators with Sufficient Monotonicity

In this chapter we will illustrate the use of Theorem 11.13 (correspondence principle), which represents the part of the theory in which there is an equivalence between comparison at the potential-theoretic (subequation $\mathcal{F}$) level and the PDE (operator F) level. We will illustrate cases in which Theorem 11.13 is applicable and cases in which it is not, and give some indication as to how one might proceed in cases *not* covered herein.

More precisely, we present comparison principles on bounded domains $\Omega \subset \mathbb{R}^n$ for constant-coefficient proper elliptic nonlinear partial differential equations

$$F(u, Du, D^2u) = c \quad \text{in } \Omega, \tag{12.1}$$

where the nonlinear operator defined by F has an associated *compatible subequation constraint set* $\mathcal{F}$ in the sense of Definition 11.1. Comparison will be formulated for $\mathcal{F}$-*admissible viscosity subsolutions and supersolutions* of (12.1) in the sense of Definition 11.12, which means that, for each $x_0 \in \Omega$,

$$J \in J^{2,+}_{x_0} u \implies J \in \mathcal{F} \text{ and } F(J) \geq c \tag{12.2}$$

and

$$J \in J^{2,-}_{x_0} u \implies \text{either } [J \in \mathcal{F} \text{ and } F(J) \leq c] \text{ or } J \notin \mathcal{F}, \tag{12.3}$$

respectively, where $J^{2,\pm}_{x_0}$ are the spaces of upper (lower) 2-jets for u at x_0.

We will require that F is *topologically tame* in the sense of Definition 11.9 and will examine structural conditions on F for which the pair $(F, \mathcal{F})$ is $\mathcal{M}$-monotone for some monotonicity cone subequation in the sense of Definition 11.2. Under these assumptions, Theorem 11.13 states that for each $c \in \mathbb{R}$ and each domain Ω, comparison holds at the PDE level for the equation $F = c$ on Ω if and only if comparison holds on Ω in the potential-theoretic sense for the subequation constraint set

$$\mathcal{F}_c := \{J \in \mathcal{F} : F(J) \geq c\}. \tag{12.4}$$

12.1   PROPER ELLIPTIC GRADIENT-FREE OPERATORS

We begin with a class of compatible pairs with monotonicity cone $\mathcal{M}(\mathcal{N}, \mathcal{P}) := \mathcal{N} \times \mathbb{R}^n \times \mathcal{P}$.

Definition 12.1. A pair $(F, \mathcal{F})$ is called a *compatible proper elliptic gradient-free operator–subequation pair* if

$$F(r, p, A) := G(r, A) \quad \text{and} \quad \mathcal{F} := \{(r, p, A) \in \mathcal{J}^2 : (r, A) \in \mathcal{G}\}, \qquad (12.5)$$

where $G \colon \mathcal{G} \subseteq \mathbb{R} \times \mathcal{S}(n) \to \mathbb{R}$ is continuous and such that the pair $(G, \mathcal{G})$ satisfies the following conditions: $\mathcal{G}$ is closed, nonempty, the pair is $\mathcal{Q}$-monotone, that is,

$$\mathcal{G} + \mathcal{Q} \subset \mathcal{G}, \quad \text{where } \mathcal{Q} = \mathcal{N} \times \mathcal{P} = \{(s, P) \in \mathbb{R} \times \mathcal{S}(n) : s \leq 0 \text{ and } P \geq 0\}, \quad (12.6)$$

and

$$G(r + s, A + P) \geq G(r, A) \quad \text{for each } (r, A) \in \mathcal{G} \text{ and each } (s, P) \in \mathcal{Q}, \qquad (12.7)$$

and finally either $\mathcal{G} = \mathbb{R} \times \mathcal{S}(n)$ (the unconstrained case) or

$$c_0 := \inf_{\mathcal{G}} G \text{ is finite} \quad \text{and} \quad \partial \mathcal{G} = \{(r, A) \in \mathcal{G} : G(r, A) = c_0\} \qquad (12.8)$$

(the constrained case).

Under hypotheses (12.6)–(12.8), it is clear that the pair $(F, \mathcal{F})$ is indeed a compatible proper elliptic operator–subequation pair in the sense of Definitions 11.1 and 11.6. One could, of course, suppress the variable $p \in \mathbb{R}^n$ and merely consider the reduced pair $(G, \mathcal{G})$ with the reduced monotonicity cone $\mathcal{M}'(\mathcal{N}, \mathcal{P}) = \mathcal{N} \times \mathcal{P}$.

Note that (12.6)–(12.7) say that $(F, \mathcal{F})$ is $\mathcal{M} = \mathcal{M}(\mathcal{N}, \mathcal{P})$-monotone in the sense of Definition 11.2. Lemma 11.3 then applies so that this hypothesis is equivalent to the statement that, for every $c \in \mathbb{R}$, the upper level set

$$\mathcal{F}_c := \{(r, p, A) \in \mathcal{F} : F(r, p, A) = G(r, A) \geq c\} \qquad (12.9)$$

is a subequation constraint set which is $\mathcal{M}(\mathcal{N}, \mathcal{P})$-monotone.

The comparison principle in this situation is now a restatement of the previously developed theory.

Theorem 12.2. *Suppose that $(F, \mathcal{F})$ is a compatible proper elliptic gradient-free pair as in Definition 12.1. Then, for every bounded domain Ω and every $c \in F(\mathcal{F})$, one has the comparison principle*

$$u \leq w \text{ on } \partial\Omega \;\; \Rightarrow \;\; u \leq w \text{ on } \Omega \qquad (12.10)$$

for each pair $u \in \mathrm{USC}(\overline{\Omega})$ *and* $w \in \mathrm{LSC}(\overline{\Omega})$ *with*

(a) u *is* $\mathcal{F}_c$-*subharmonic and* w *is* $\mathcal{F}_c$-*superharmonic (that is,* $-w$ *is* $\widetilde{\mathcal{F}}_c$-*subharmonic).*

If one also requires the additional hypothesis that the operator F *is topologically tame, that is,*

$$\{(r, p, A) \in \mathcal{F} : F(r, p, A) = G(r, A) = c\} \text{ has empty interior for all } c \in \mathbb{R},$$
$$(12.11)$$

then one can replace (a) *by the equivalent hypothesis that* $u \in \mathrm{USC}(\overline{\Omega})$ *and* $w \in \mathrm{LSC}(\overline{\Omega})$ *satisfy*

(b) u *is an* $\mathcal{F}$-*admissible viscosity subsolution and* w *is an* $\mathcal{F}$-*admissible supersolution to* $F(u, Du, D^2 u) := G(u, D^2 u) = c$ *on* Ω.

Here, as always, $c \in F(\mathcal{F})$.

Proof. The comparison principle (12.10) for $\mathcal{F}_c$-subharmonics u and superharmonics w, on every bounded domain Ω, follows from Theorem 10.3 since $\mathcal{F}_c$ is a gradient-free subequation. Finally, by Theorem 11.13, since F is topologically tame and since $(F, \mathcal{F})$ is a compatible proper elliptic operator–subequation pair which is $\mathcal{M}$-monotone for a convex cone subequation, for each $c \in F(\mathcal{F})$, the comparison principle holds for $\mathcal{F}$-admissible viscosity subsolutions u and supersolutions w of $F(u, Du, D^2 u) = G(u, D^2 u) = c$. $\square$

We remark that Theorem 12.2 also includes the pure second-order case where $F(r, p, A) := G(A)$, with G increasing on a closed nonempty $\mathcal{P}$-monotone subset $\mathcal{G}$ of $\mathcal{S}(n)$ such that $(G, \mathcal{G})$ is a compatible pair (with the obvious reduction of also suppressing the variable $r \in \mathbb{R}$ and using the reduced monotonicity cone $\mathcal{M}'(\mathcal{P}) = \mathcal{P}$).

We conclude this section with some examples to illustrate the applicability of Theorem 12.2, as well as some "bad" examples for which Theorem 12.2 does not apply in this topologically tame proper elliptic gradient-free pairs case. The first examples make use of *canonical operators* in *unconstrained cases.*

Example 12.3 (Gradient-free canonical operators). Consider any gradient-free subequation $\mathcal{F} \subsetneq \mathcal{J}^2$ (see Definition 10.2), that is, $\mathcal{F}$ is closed, nonempty, and $\mathcal{M}(\mathcal{N}, \mathcal{P}) := \mathcal{N} \times \mathbb{R}^n \times \mathcal{P}$-monotone. Fix $J_0 \in \mathrm{Int}\, \mathcal{M}(\mathcal{N}, \mathcal{P})$ (which the reader may wish to standardize as $J_0 = (-1, 0, I)$) and let $F \in C(\mathcal{J}^2)$ be the canonical operator for $\mathcal{F}$ (see Definition 11.16); that is, for each $J \in \mathcal{J}^2$,

$$F(J) := -t_J, \quad \text{where } t_J \text{ is the unique element of } \mathbb{R} \text{ such that } J + t_J J_0 \in \partial \mathcal{F}.$$

Since $\mathcal{M} = \mathcal{M}(\mathcal{N}, \mathcal{P})$ is a monotonicity cone subequation, by Theorem 11.20, both $(F, \mathcal{J}^2)$ and the dual pair $(\widetilde{F}, \mathcal{J}^2)$ are unconstrained compatible proper elliptic operator–subequation pairs with F and $\widetilde{F}$ topologically tame. Hence, the comparison principle of Theorem 12.2 applies to both pairs.

It is worth stressing that each gradient-free subequation $\mathcal{F}$ gives rise to a family of admissible operators (parameterized by $J_0 \in \mathrm{Int}\,\mathcal{M}(\mathcal{N}, \mathcal{P})$). With the standard choice $J_0 = (-1, 0, I)$, the canonical operator is $F(J) = F(r, p, A) = \min\{-r, \lambda_{\min}(A)\}$ by (11.60) and with dual $\widetilde{F}(J) := -F(-J) = \max\{-r, \lambda_{\max}(A)\}$ by (11.46).

Example 12.4 (Perturbations of pure second-order canonical operators). Consider any pure second-order subequation $\mathcal{H} \subset \mathcal{S}(n)$, that is, $\mathcal{H}$ is closed, proper, nonempty, and $\mathcal{P}$-monotone. Define a gradient-free operator $F\colon \mathcal{J}^2 \to \mathbb{R}$ by

$$F(r, p, A) := H(A) + h(r), \qquad (12.12)$$

where $h \in C(\mathbb{R})$ is nonincreasing and $H \in C(\mathcal{S}(n))$ is the canonical operator for $\mathcal{H}$ (determined by $A_0 \in \mathrm{Int}\,\mathcal{P}$, where the standard choice is $A_0 := I$). This operator is given by the pure second-order version of Definition 11.16, that is, for each $A \in \mathcal{S}(n)$,

$$H(A) := -t_A, \quad \text{where } t_A \text{ is the unique element of } \mathbb{R} \text{ such that } A + t_A A_0 \in \partial \mathcal{H},$$

and one has

$$H(A + t A_0) = H(A) + t \quad \text{for each } A \in \mathcal{S}(n),\ t \in \mathbb{R}. \qquad (12.13)$$

One has that $(F, \mathcal{J}^2)$ is an unconstrained operator–subequation pair which is $\mathcal{M}(\mathcal{N}, \mathcal{P})$-monotone since $F(r, p, A)$ is increasing in A by (12.13) and nonincreasing in r by the monotonicity hypothesis on h. Moreover, for each $t > 0$ (in the case that J_0 is the standard choice for simplicity),

$$\begin{aligned} F((r, p, A) + t(-1, 0, I)) &= H(A) + t + h(r - t) \\ &\geq H(A) + t + h(r) \\ &= F(r, p, A) + t, \end{aligned}$$

so that F is topologically tame by Theorem 11.10(3). Hence, the comparison principle of Theorem 12.2 applies to $(F, \mathcal{J}^2)$.

One could also use the dual operator $\widetilde{H}$ for the dual subequation $\widetilde{\mathcal{H}}$. More generally, one can take finite sums

$$F(r, p, A) = \sum_{k=1}^{N} H_k(A) + h(r)$$

if each $\mathcal{H}_k \subsetneq \mathcal{S}(n)$ is a pure second-order subequation with canonical operator H_k (determined by the same $A_0 \in \mathrm{Int}\,\mathcal{P}$).

A simple instructive example is $F(r, p, A) := \lambda_{\min}(A) - r$, which is the sum of the canonical operators for $\mathcal{M}(\mathcal{P})$ and $\mathcal{M}(\mathcal{N})$.

Next we turn to the constrained gradient-free case, where *Gårding–Dirichlet polynomials* generate many interesting examples. The reader might want to look back at Section 11.6 (especially Examples 11.33 and 11.34) to review just how rich the class of examples is.

Example 12.5 (Operators involving Gårding–Dirichlet polynomials). Consider $G(r, A) := h(r)\mathfrak{g}(A)$, where $\mathfrak{g}$ is a Gårding–Dirichlet polynomial of degree m in the sense of Definition 11.31 and $h \in C((-\infty, 0])$ is continuous with

$$h \text{ is nonincreasing, } h \geq 0, \text{ and } h(0) = 0 \iff r = 0. \tag{12.14}$$

Take as the domain for G the set $\mathcal{G} := \mathcal{N} \times \overline{\Gamma} \subset \mathbb{R} \times \mathcal{S}(n)$, where $\mathcal{N} = (-\infty, 0]$ and $\overline{\Gamma} = \{A \in \mathcal{S}(n) : \lambda_k^{\mathfrak{g}}(A) \geq 0, \ k = 1, \ldots, n\}$ is the closed Gårding cone. The cone $\overline{\Gamma}$ is assumed to satisfy $\overline{\Gamma} + \mathcal{P} \subset \overline{\Gamma}$, or equivalently $\mathcal{P} \subset \overline{\Gamma}$. Recall that by normalizing $\mathfrak{g}(A)$ (which is I-hyperbolic) to have $\mathfrak{g}(I) = 1$, one has

$$\mathfrak{g}(A + tI) = \prod_{k=1}^{m}(\lambda_k^{\mathfrak{g}}(A) + t) \quad \text{and} \quad \mathfrak{g}(A) = \prod_{k=1}^{m} \lambda_k^{\mathfrak{g}}(A). \tag{12.15}$$

Since $h \geq 0$ on $\mathcal{N}$ and since $\lambda_k^{\mathfrak{g}}(A) \geq 0$ for each k defines $\overline{\Gamma}$, one has that

$$G(r, A) = h(r)\mathfrak{g}(A) \geq 0 \quad \text{for each } (r, A) \in \mathcal{G} := \mathcal{N} \times \overline{\Gamma},$$

and since $G(0, A) = 0$, one has the first compatibility condition

$$\inf_{\mathcal{G}} G = 0 \quad \text{(finite)}.$$

Moreover, using the third property in (12.14) one easily verifies the second compatibility condition

$$\partial\mathcal{G} = \{(r, A) \in \mathcal{G} : G(r, A) = h(r)\mathfrak{g}(A) = 0\}.$$

Thus $(G, \mathcal{G})$ is a constrained-case compatible gradient-free operator–subequation pair. Also, G is proper elliptic and topologically tame since for each $(r, A) \in \mathcal{G}$ and each $(s, P) \in \mathcal{Q} = \mathcal{N} \times \mathcal{P}$, one has

$$G(r + s, A + P) = h(r + s)\mathfrak{g}(A + P) \geq h(r)\mathfrak{g}(A + P) \geq h(r)\mathfrak{g}(A),$$

and with $J_0' = (-1, I) \in \text{Int } \mathcal{Q}$ and $t > 0$ one has

$$G((r,A)+t(-1,I)) = h(r-t)\mathfrak{g}(A+tI) = h(r-t)\prod_{k=1}^{m}(\lambda_k^{\mathfrak{g}}(A)+t)$$

$$\geq h(r)\prod_{k=1}^{m}(\lambda_k^{\mathfrak{g}}(A)+t) \geq h(r)\prod_{k=1}^{m}\lambda_k^{\mathfrak{g}}(A) = G(r,A).$$

Hence, the comparison theorem (Theorem 12.2) applies to the constrained case pair $(G,\mathcal{G})$.

Obviously, one could replace $\mathfrak{g}(A)$ and $\overline{\Gamma}$ by any of its factors $\lambda_k^{\mathfrak{g}}(A)$ and the branch $\Lambda_k^{\mathfrak{g}} := \{A \in \mathcal{S}(n) : \lambda_k^{\mathfrak{g}}(A) \geq 0\}$. Moreover, the same holds for $G(r,A) = h(r)H(A)$ if h is above and $H(A)$ is the canonical operator (determined by $J_0 \in \operatorname{Int}\mathcal{P}$) for $\mathcal{H} \subset \mathcal{S}(n)$ a $\mathcal{P}$-invariant pure second-order subequation.

Example 12.6 (The hyperbolic affine sphere equation). The partial differential equation

$$\det(D^2u) = \left(\frac{L}{u}\right)^{n+2}, \quad L \leq 0 \tag{12.16}$$

arises in the study of *hyperbolic affine spheres* with mean curvature L, where $u < 0$ is convex and vanishes on the boundary of $\Omega \subset \mathbb{R}^n$ convex (see Cheng–Yau [22] and the references therein). Equation (12.16) is covered by Example 12.5 if one takes $\mathfrak{g}(A) = \det(A)$ and $h(r) = (-r)^{n+2}$, and $-L \geq 0$ corresponds to the admissible levels.

Remark 12.7. Products like $G(r,A) = -r\det(A)$ in Example 12.5 (where $\mathfrak{g}(A) = \det(A)$ and $h(r) = -r$) are good examples which are admissible for the correspondence principle (Theorem 11.13) in the constrained case. However, sums like

$$G(r,A) := \det(A) - r \tag{12.17}$$

are bad examples. For this example,

$$G(r+s, A+P) \geq G(r,A) \quad \text{for all } (r,A) \in \mathbb{R} \times \mathcal{P}, \ (s,P) \in \mathcal{N} \times \mathcal{P},$$

so that, with domain $\mathcal{G} = \mathbb{R} \times \mathcal{P}$, the pair $(G,\mathcal{G})$ is a $\mathcal{Q} = \mathcal{N} \times \mathcal{P}$-monotone gradient-free pair. However, with this maximal domain $\mathbb{R} \times \mathcal{P}$ of $\mathcal{Q}$-monotonicity of the operator G,

$$\inf_{\mathcal{G}} G = -\infty \quad \text{(is not finite)} \tag{12.18}$$

and hence the pair fails to satisfy the first condition in (12.8) for compatibility. If one cuts down the domain to $\mathcal{G} := (-\infty, 0] \times \mathcal{P}$, so that the inf in (12.18) is zero (finite), then the jet subset

$$\mathcal{F} \equiv \{(r,A) \in (-\infty, 0] \times \mathcal{P} : \det(A) - r \geq 0\} \tag{12.19}$$

has boundary $\partial \mathcal{F}$ including $(-\infty, 0] \times \{0\}$, so that all negative C^2-affine functions are $\mathcal{F}$-harmonic but the operator G is not zero. Since $\partial \mathcal{F}$ is much larger than the zero set of the operator $\{(r, A) \in \mathcal{G} : G(r, A) = 0\}$, the second condition in (12.8) for compatibility fails.

The problem for these examples is that the subharmonics of $-\widetilde{\mathcal{G}}$ do not correspond to $\mathcal{G}$-admissible supersolutions of the equation $\mathcal{H}$ defined by $G(r, A) = 0$. However, for such examples, one can make use of the notion of a *generalized equation* in which one looks for another subequation constraint set $\mathcal{E} \subset \mathbb{R} \times \mathcal{S}(n)$ (different from $\mathcal{G}$) such that

$$\mathcal{H} = \mathcal{G} \cap (-\widetilde{\mathcal{E}}).$$

We will not pursue this further here, but refer to [61] for details, where the pure second-order case is discussed at length.

In order to treat situations with gradient dependence, as is commonly known, a Lipschitz condition in p is helpful. These ideas will be explored next. The next three sections treat unconstrained compatible pairs $(F, \mathcal{J}^2)$, which also indicate how many classical results can be recovered by our monotonicity method in the presence of a suitable monotonicity cone $\mathcal{M}$ for the pair.

12.2 DEGENERATE ELLIPTIC OPERATORS WITH STRICT MONOTONICITY IN r

Our next result concerns the unconstrained case of a well-known example class of operators. It also shows how the $\mathcal{M}(\gamma, \mathbb{R}^n, \mathcal{P})$-cones with $\gamma \geq 0$ arise naturally in an important example class with gradient dependence.

Theorem 12.8. *Suppose that $F: \mathcal{J}^2 \to \mathbb{R}$ is continuous and satisfies the following structural condition: there exist $\alpha > 0$ and $\beta \geq 0$ such that, for each $(r, p, A) \in \mathcal{J}^2$ and each $(s, q, P) \in \mathcal{N} \times \mathbb{R}^n \times \mathcal{P}$, one has*

$$F(r + s, p + q, A + P) - F(r, p, A) \geq -\alpha s - \beta |q|. \tag{12.20}$$

Then $(F, \mathcal{J}^2)$ is an unconstrained compatible proper elliptic operator–subequation pair which is $\mathcal{M}$-monotone for the monotonicity cone subequation

$$\mathcal{M} = \mathcal{M}(\gamma, \mathbb{R}^n, \mathcal{P}) := \{(s, q, P) \in \mathcal{J}^2 : s \leq -\gamma |q|, \ q \in \mathbb{R}^n, \ P \in \mathcal{P}\} \tag{12.21}$$

if $\gamma := \beta / \alpha$. Consequently, for each admissible level $c \in F(\mathcal{J}^2)$, the set

$$\mathcal{F}_c := \{(r, p, A) \in \mathcal{J}^2 : F(r, p, A) \geq c\} \tag{12.22}$$

is $\mathcal{M}(\gamma, \mathbb{R}^n, \mathcal{P})$-*monotone. Moreover, the operator* F *is topologically tame, and hence for every bounded domain* Ω *one has the comparison principle*

$$u \leq w \text{ on } \partial\Omega \quad \Rightarrow \quad u \leq w \text{ on } \Omega \qquad (12.23)$$

for $u \in \mathrm{USC}(\overline{\Omega})$ *and* $w \in \mathrm{LSC}(\overline{\Omega})$, *which are respectively* $\mathcal{F}_c$-*subharmonic and* $\mathcal{F}_c$-*superharmonic in* Ω, *or equivalently, if* u *and* w *are respectively viscosity subsolutions and supersolutions to* $F(u, Du, D^2u) = c$ *on* Ω.

Proof. By definition (11.6) and the hypothesis $F \in C(\mathcal{J}^2)$, with $\mathcal{F} := \mathcal{J}^2$, one has that $(F, \mathcal{F})$ is an unconstrained compatible pair, and $\mathcal{F} = \mathcal{J}^2$ is trivially $\mathcal{M}$-monotone for every $\mathcal{M}$. Using the structural condition (12.20), one has for each $(r, p, A) \in \mathcal{J}^2$ and each $(s, q, P) \in \mathcal{M}(\gamma, \mathbb{R}^n, \mathcal{P}) \subset \mathcal{N} \times \mathbb{R}^n \times \mathcal{P}$,

$$F(r+s, p+q, A+P) \geq F(r, p, A) - \alpha s - \beta|q| \geq F(r, p, A), \qquad (12.24)$$

where $-\alpha s - \beta|q| \geq 0$ since $s \leq -\frac{\beta}{\alpha}|q|$. Hence, F is $\mathcal{M}(\gamma, \mathbb{R}^n, \mathcal{P})$-monotone and since $\mathcal{M}(\gamma, \mathbb{R}^n, \mathcal{P}) \supset \mathcal{M}_0 = \mathcal{N} \times \{0\} \times \mathcal{P}$, the pair is proper elliptic in accordance with Definition 11.6. By Lemma 11.3, every upper level set $\mathcal{F}_c$ is $\mathcal{M}(\gamma, \mathbb{R}^n, \mathcal{P})$-monotone and the comparison (12.23) on each bounded domain Ω for $\mathcal{F}_c$-subharmonic, $\mathcal{F}_c$-superharmonic pairs follows from Theorem 7.6 by the $\mathcal{M}(\gamma, \mathbb{R}^n, \mathcal{P})$-monotonicity of each $\mathcal{F}_c$.

Finally, F is topologically tame by Theorem 11.10(2) since $F(J + J_0) > F(J)$ for each $J \in \mathcal{J}^2$ and $J_0 \in \mathrm{Int}\,\mathcal{M}(\gamma, \mathbb{R}^n, \mathcal{P})$, where $-\alpha s - \beta|q| > 0$ in (12.24). Hence, the comparison principle (12.23) for (unconstrained) viscosity subsolution, supersolution pairs (u, v) of $F(u, Du, D^2u) = c$ is equivalent to the comparison for $\mathcal{F}_c$-subharmonic, $\mathcal{F}_c$-superharmonic pairs by Theorem 11.13. $\qquad \square$

Concerning Theorem 12.8, a few examples and remarks are worth noting.

Example 12.9. The case $\beta = 0$ of Theorem 12.8 yields comparison for proper elliptic gradient-free operators $F(r, p, A) = G(r, A)$ with strict monotonicity in the r-variable, that is, $G \in C(\mathbb{R} \times \mathcal{S}(N))$ such that, for some $\alpha > 0$, one has

$$G(r+s, A+P) - G(r, A) \geq -\alpha s \quad \forall\, (r, A) \in \mathbb{R} \times \mathcal{S}(N), \; s \leq 0, \; P \geq 0. \quad (12.25)$$

With respect to the unconstrained case of Theorem 12.2, the strict monotonicity in r ensures the topological tameness, which was a hypothesis in the previous theorem.

The case $\beta > 0$ of Theorem 12.8 yields comparison for operators of the form

$$F(r, p, A) = G(r, A) + \langle b, p \rangle, \qquad (12.26)$$

with G as above and $b \in \mathbb{R}^n \setminus \{0\}$. One can then choose $\beta = |b|$. Notice that if the strict monotonicity constant $\alpha = 1$ in (12.25), then $\gamma = \beta$, that is, the monotonicity cone parameter equals $|b|$.

Remark 12.10. Theorem 12.8 is closely related to one of the two cases in the groundbreaking paper of Jensen [69] on the maximum principle for viscosity solutions to constant-coefficient equations. In the situation of degenerate ellipticity and strict monotonicity in r, the first case of Jensen's Theorem 3.1 in [69] gives a maximum principle (which then implies the comparison principle) for a subsolution/supersolution pair with more regularity than we require. He assumes that the pair belongs to $C(\overline{\Omega}) \cap W^{1,\infty}(\Omega)$. On the other hand, Jensen does not require the Lipschitz in p condition which we need in order to have a monotonicity cone with nonempty interior. The second case of Jensen's theorem (for uniformly elliptic operators which are Lipschitz in p) will be discussed in the next section (see Theorem 12.16).

12.3 PROPER OPERATORS WITH SOME DEGREE OF STRICT ELLIPTICITY

In this section we examine classes of proper operators with gradient dependence in which the weak monotonicity assumption of degenerate ellipticity is strengthened to include some measure of strict monotonicity in the Hessian variable, but there may be no strict monotonicity in the r-variable as in the previous section. Our first result is stated in the unconstrained case, as was done in Theorem 12.8. It makes use of $\mathcal{M}(\mathcal{N}, R)$-monotonicity with finite R.

Theorem 12.11. *Suppose that $F \colon \mathcal{J}^2 \to \mathbb{R}$ is continuous and is proper elliptic; that is, it satisfies for each $(r, p, A) \in \mathcal{J}^2$ and each $(s, P) \in \mathcal{N} \times \mathcal{P}$,*

$$F(r + s, p, A + P) \geq F(r, p, A). \qquad (12.27)$$

In addition, suppose that F satisfies the following structural condition: there exist $\alpha, \beta > 0$ such that, for each $(r, p, A) \in \mathcal{J}^2$ and each $\mu \geq 0, q \in \mathbb{R}^n$, one has

$$F(r, p + q, A + \mu I) - F(r, p, A) \geq \alpha \mu - \beta |q|. \qquad (12.28)$$

Then $(F, \mathcal{J}^2)$ is an unconstrained proper elliptic compatible operator–subequation pair which is $\mathcal{M}$-monotone for the monotonicity cone subequation

$$\mathcal{M} = \mathcal{M}(\mathcal{N}, R) := \left\{ (s, q, P) \in \mathcal{J}^2 : s \leq 0, \ q \in \mathbb{R}^n, \text{ and } P \geq \tfrac{|q|}{R} I \right\} \qquad (12.29)$$

if $R \leq \alpha/\beta$. Consequently, if $R \leq \alpha/\beta$ then for each $c \in F(\mathcal{J}^2)$ the set

$$\mathcal{F}_c := \left\{ (r, p, A) \in \mathcal{J}^2 : F(r, p, A) \geq c \right\} \qquad (12.30)$$

is $\mathcal{M}(\mathcal{N}, R)$-monotone. Moreover, the operator F is topologically tame, and hence for every bounded domain Ω which is contained in a translate of $B_R(0)$, *one has the comparison principle*

$$u \leq w \ on \ \partial\Omega \ \Rightarrow \ u \leq w \ on \ \Omega \tag{12.31}$$

for $u \in \mathrm{USC}(\overline{\Omega})$ and $w \in \mathrm{LSC}(\overline{\Omega})$, which are respectively $\mathcal{F}_c$-subharmonic and $\mathcal{F}_c$-superharmonic in Ω, or equivalently, if u and w are respectively viscosity subsolutions and supersolutions to $F(u, Du, D^2u) = c$ on Ω.

Proof. One follows the same argument as the proof of Theorem 12.8. It is clear that $(F, \mathcal{J}^2)$ is an unconstrained proper elliptic operator–subequation pair and that $\mathcal{J}^2$ is trivially $\mathcal{M}$-monotone. For the $\mathcal{M}$-monotonicity of F, if $(r, p, A) \in \mathcal{J}^2$ and $(s, q, P) \in \mathcal{M}(\mathcal{N}, R)$ then by (12.27)–(12.29) one has

$$F(r + s, p + q, A + P) \geq F\left(r, p + q, A + \frac{|q|}{R}I\right)$$

$$\geq F(r, p, A) + |q|\left(\frac{\alpha}{R} - \beta\right) \geq 0,$$

where the condition $R \leq \alpha/\beta$ is needed. Again, by Lemma 11.3, every upper level set $\mathcal{F}_c$ is $\mathcal{M}(\mathcal{N}, R)$-monotone and the comparison (12.31) on each domain Ω contained in a translate of $B_R(0)$ for $\mathcal{F}_c$-subharmonic, superharmonic pairs follows from Theorem 7.6 by the $\mathcal{M}(\mathcal{N}, R)$-monotonicity of each $\mathcal{F}_c$. The topological tameness of F again follows from Theorem 11.10(2) by using the strict monotonicity in (12.28). Hence, by Theorem 11.13, the comparison (12.31) also holds for viscosity subsolution, supersolution pairs (u, w) of $F(u, Du, D^2u) = c$ for each $c \in \mathbb{R}$ and each Ω contained in a translate of $B_R(0)$. $\qquad\square$

Before proceeding, we make a remark on the terminology of notions of ellipticity.

Remark 12.12. The monotonicity property in A of (12.28), that is, with $\alpha > 0$,

$$F(r, p, A + \mu I) - F(r, p, A) \geq \alpha\mu \quad \text{for each } \mu \geq 0, \tag{12.32}$$

might well be called *strict (uniform) ellipticity in the direction $I \in \mathcal{S}(n)$*. In Bardi–Mannucci [5] this partial (strict) uniform ellipticity was called *nontotally degenerate ellipticity* since in the quasi-linear case it corresponds to what Bony called nontotally degenerate in [11]. One should also note that in the language of [61], condition (12.32) would be called *(linear) tameness in A of F on $\mathcal{J}^2$*.

It is worth noting the limit case of $\beta = 0$ in Theorem 12.11, which gives a gradient-free situation with $\mathcal{M}(\mathcal{N}, \mathcal{P})$-monotonicity and comparison on arbitrary bounded domains.

Theorem 12.13. *Suppose that* $F\colon \mathcal{J}^2 \to \mathbb{R}$ *is continuous and is proper elliptic, that is, it satisfies for each* $(r, p, A) \in \mathcal{J}^2$ *and each* $(s, P) \in \mathcal{N} \times \mathcal{P}$,

$$F(r + s, p, A + P) \geq F(r, p, A). \tag{12.33}$$

In addition, suppose that F *satisfies the following structural condition: there exist* $\alpha > 0$ *such that, for each* $(r, p, A) \in \mathcal{J}^2$ *and each* $\mu \geq 0$, $q \in \mathbb{R}^n$, *one has*

$$F(r, p + q, A + \mu I) - F(r, p, A) \geq \alpha \mu. \tag{12.34}$$

Then $(F, \mathcal{J}^2)$ *is an unconstrained proper elliptic compatible operator–subequation pair which is gradient-free and* $\mathcal{M}$-*monotone for the monotonicity cone subequation*

$$\mathcal{M} = \mathcal{M}(\mathcal{N}, \mathcal{P}) := \left\{ (s, q, P) \in \mathcal{J}^2 : s \leq 0, \ q \in \mathbb{R}^n, \ and \ P \geq 0 \right\}. \tag{12.35}$$

Consequently, for each $c \in F(\mathcal{J}^2)$ *the set*

$$\mathcal{F}_c := \left\{ (r, p, A) \in \mathcal{J}^2 : F(r, p, A) \geq c \right\} \tag{12.36}$$

is $\mathcal{M}(\mathcal{N}, \mathcal{P})$-*monotone. Moreover, the operator* F *is topologically tame and hence for every bounded domain* Ω, *one has the comparison principle*

$$u \leq w \ on \ \partial\Omega \ \Rightarrow \ u \leq w \ on \ \Omega \tag{12.37}$$

for $u \in \mathrm{USC}(\overline{\Omega})$ *and* $w \in \mathrm{LSC}(\overline{\Omega})$, *which are respectively* $\mathcal{F}_c$-*subharmonic and* $\mathcal{F}_c$-*superharmonic in* Ω, *or equivalently, if* u *and* w *are respectively viscosity subsolutions and supersolutions to* $F(u, Du, D^2 u) = c$ *on* Ω.

Proof. The proof is almost identical to that of Theorem 12.11. We limit ourselves to showing that the pair $(F, \mathcal{J}^2)$ is gradient-free, which is equivalent to the $\mathcal{M}(\mathcal{N}, \mathcal{P})$-monotonicity of F on all of $\mathcal{J}^2$. By combining (12.33) with (12.34) with $\mu = 0$, for each $(r, p, A) \in \mathcal{J}^2$ and each $(s, q, P) \in \mathcal{M}(\mathcal{N}, \mathcal{P})$, one has

$$F(r + s, p + q, A + P) \geq F(r, p + q, A) = F(r, p + q, A + 0I) \geq F(r, p, A) + 0,$$

as needed. $\square$

Next we give a few examples covered by Theorems 12.11 and 12.13.

Example 12.14. The case $\beta = 0$ of Theorem 12.13 yields comparison on arbitrary bounded domains for proper elliptic gradient-free operators $F(r, p, A) = G(r, A)$ which are strictly elliptic in the direction $I \in \mathcal{S}(N)$, that is, $G \in C(\mathbb{R} \times \mathcal{S}(N))$ such that

$$G(r+s, A+P) \geq G(r, A) \quad \forall\, (r, A) \in \mathbb{R} \times \mathcal{S}(N), \ s \leq 0, \ P \geq 0, \qquad (12.38)$$

and for some $\alpha > 0$,

$$G(r, A+\mu I) - G(r, A) \geq \alpha \mu \quad \forall\, (r, A) \in \mathbb{R} \times \mathcal{S}(N), \ \mu \geq 0. \qquad (12.39)$$

With respect to the unconstrained case of Theorem 12.2, the strict monotonicity in (12.39) ensures the topological tameness, which was a hypothesis in the previous theorem.

The case $\beta > 0$ of Theorem 12.11 yields comparison on domains contained in translates of $B_R(0)$ with $R \leq \frac{\alpha}{\beta}$ for operators of the form

$$F(r, p, A) = G(r, A) - \langle b, p \rangle, \qquad (12.40)$$

with G as above and $b \in \mathbb{R}^n \setminus \{0\}$. One can then choose $\beta = |b|$.

The next remark concerns the result obtained in Theorem 12.11.

Remark 12.15. The restriction that $R \leq \frac{\alpha}{\beta}$ can be viewed in two ways. First, one can say that comparison holds on all bounded domains Ω provided that the Lipschitz constant β is small relative to the diameter $2R$ of Ω and the partial ellipticity constant α, that is, if

$$\beta \leq \frac{\alpha}{R}. \qquad (12.41)$$

On the other hand, for fixed Lipschitz constant β and partial ellipticity constant α, comparison is ensured only for domains with diameter $2R$ satisfying

$$R \leq \frac{\alpha}{\beta}. \qquad (12.42)$$

This remark raises two interesting questions. The first question is what minimal further strengthening of the notion of ellipticity gives comparison on arbitrarily large domains independent of the Lipschitz constant β? One classical answer will be given below in the constrained context by strengthening the notion of ellipticity. The second question is whether the monotonicity cone $\mathcal{M}(\mathcal{N}, \mathcal{D}, R)$ is the maximal monotonicity cone $\mathcal{M}_{\mathcal{F}}$ for $\mathcal{F}$ defined by (12.30). Perhaps, at least in some important special cases, additional structure in F can lead to a larger monotonicity cone with strict approximators (and hence comparison) on arbitrary domains with large Lipschitz constant. This second question was considered at the potential-theoretic level in Chapter 8.

The following strengthening of the ellipticity recovers Jensen's uniformly elliptic result[1] in [69] by using our monotonicity method and adds an unconstrained potential-theoretic version of the result at all admissible levels $c \in F(\mathcal{J}^2)$.

---

[1]We will use the term *strictly elliptic* since we are asking only for a one-sided ellipticity bound, reserving *uniformly elliptic* for a two-sided bound.

Theorem 12.16. *Suppose that $F\colon \mathcal{J}^2 \to \mathbb{R}$ is continuous and that F is proper and strictly elliptic; that is, there exists $\lambda > 0$ such that, for each $(r, p, A) \in \mathcal{J}^2$, one has*

$$F(r+s, p, A+P) - F(r, p, A) \geq \lambda \operatorname{tr} P \quad \text{for each } s \in \mathcal{N} \text{ and each } P \in \mathcal{P}.$$
$$(12.43)$$

In addition, suppose that F satisfies the following structural condition: there exists $\beta > 0$ such that, for each $(r, p, A) \in \mathcal{J}^2$, one has

$$F(r, p+q, A) - F(r, p, A) \geq -\beta|q| \quad \text{for each } q \in \mathbb{R}^n. \qquad (12.44)$$

Then $(F, \mathcal{J}^2)$ is an unconstrained proper elliptic compatible operator–subequation pair which is $\mathcal{M}$-monotone for the monotonicity cone subequation

$$\mathcal{M} = \mathcal{M}_{\lambda,\beta}(\mathcal{N}, \mathcal{D}, \mathcal{P}) := \{(s, q, P) \in \mathcal{N} \times \mathcal{D} \times \mathcal{P} : \lambda \operatorname{tr} P \geq \beta|q|\}, \qquad (12.45)$$

with directional cone $\mathcal{D} = \mathbb{R}^n$. Consequently, for each $c \in F(\mathcal{J}^2)$ the set

$$\mathcal{F}_c := \{(r, p, A) \in \mathcal{J}^2 : F(r, p, A) \geq c\} \qquad (12.46)$$

is $\mathcal{M}_{\lambda,\beta}(\mathcal{N}, \mathbb{R}^n, \mathcal{P})$-monotone. Moreover, the operator F is topologically tame, and hence for every bounded domain Ω one has the comparison principle

$$u \leq w \text{ on } \partial\Omega \;\Rightarrow\; u \leq w \text{ on } \Omega \qquad (12.47)$$

for $u \in \operatorname{USC}(\overline{\Omega})$ and $w \in \operatorname{LSC}(\overline{\Omega})$, which are respectively $\mathcal{F}_c$-subharmonic and $\mathcal{F}_c$-superharmonic in Ω, or equivalently, if u and w are respectively viscosity subsolutions and supersolutions to $F(u, Du, D^2u) = c$ on Ω.

Proof. For each directional cone $\mathcal{D} \subseteq \mathbb{R}^n$ (in the sense of Definition 2.2), one clearly has that $\mathcal{M}_{\lambda,\beta}(\mathcal{N}, \mathcal{D}, \mathcal{P})$ is a monotonicity cone subequation which contains $\mathcal{M}_0 = \mathcal{N} \times \{0\} \times \mathcal{P}$ and that $(F, \mathcal{J}^2)$ is an unconstrained compatible proper elliptic operator–subequation pair, by imitating the argument of Theorem 12.11 where the structural conditions (12.43)–(12.44) play the same role as (12.27)–(12.28). The conclusion that each subequation $\mathcal{F}_c$ is $\mathcal{M}_{\lambda,\beta}(\mathcal{N}, \mathbb{R}^n, \mathcal{P})$-monotone again follows from Lemma 11.3. The topological tameness of F follows from Theorem 11.10(2) by using the strict monotonicity in (12.44).

Hence, by the general comparison result of Theorem 7.5, the comparison principles for each $c \in F(\mathcal{J}^2)$ and each bounded domain Ω reduce to the question of whether the monotonicity cone subequation $\mathcal{M}_{\lambda,\beta}(\mathcal{N}, \mathbb{R}^n, \mathcal{P})$ admits a strict approximator in the sense of Definition 6.1 on a given bounded domain Ω. We will give the argument for general $\mathcal{D}$, even though here we need only the special case $\mathcal{D} = \mathbb{R}^n$. Since Ω is bounded and since the convex cone $\mathcal{D}$ has interior, by translating Ω, we can assume that $\overline{\Omega} \subset \mathcal{D}_R = \mathcal{D} \cap B_R(0)$ for some $R > 0$ and that $0 \notin \Omega$. We will look for $\psi \in C^\infty(\mathbb{R}^n \setminus \{0\})$ of the form

$$\psi(x) = \frac{1}{\mu} e^{\mu|x|} - m \quad \text{with } \mu, m > 0, \tag{12.48}$$

to be determined so that for each $x \in \Omega$, one has $(s, q, P) := (\psi(x), D\psi(x), D^2 \psi(x)) \in \text{Int } \mathcal{M}_{\lambda,\beta}(\mathcal{N}, \mathcal{D})$, where

$$\text{Int } \mathcal{M}_{\lambda,\beta}(\mathcal{N}, \mathcal{D}) = \{s < 0, \ q \in \text{Int } D, \ P > 0, \text{ and } \lambda \text{tr } P > \beta|q|\}. \tag{12.49}$$

Clearly,

$$s := \psi(x) = \frac{1}{\mu} e^{\mu|x|} - m < 0 \quad \text{if } m > \frac{1}{\mu} e^{\mu R} \tag{12.50}$$

and

$$q := D\psi(x) = e^{\mu|x|} \frac{x}{|x|} \in \text{Int } \mathcal{D} \tag{12.51}$$

since $\overline{\Omega} \subset \mathcal{D} \cap (B_R(0) \setminus \{0\})$. Moreover, as computed in example (4) of Remark 3.17, the Hessian of the radial function ψ is $P := D^2\psi(x) = \frac{1}{|x|} e^{\mu|x|} P_{x^\perp} + \mu e^{\mu|x|} P_x$ and hence the eigenvalues of P are

$$\frac{e^{\mu|x|}}{|x|} \text{ (with multiplicity } n-1) \quad \text{and} \quad \mu e^{\mu|x|} \text{ (with multiplicity 1)}. \tag{12.52}$$

Hence, $P > 0$ and one has

$$\lambda \text{tr } P - \beta|q| = e^{\mu|x|} \left[\frac{\lambda(n-1)}{|x|} + \lambda\mu - \beta \right] > 0,$$

by choosing $\mu > \beta/\lambda$.   □

12.4 FROM LINEAR OPERATORS TO HAMILTON–JACOBI–BELLMAN OPERATORS

In this section we will discuss some unconstrained situations that illustrate the use of $\mathcal{M} = \mathcal{N} \times \mathcal{D} \times \mathcal{P}$-monotonicity for linear equations and certain Hamilton–Jacobi–Bellman equations. Admittedly, in the linear case, there is nothing new here, but perhaps it is still useful to illustrate how this classical case fits into this part of the theory.

Consider the class of *proper elliptic linear operators* $F \colon \mathcal{J}^2 \to \mathbb{R}$, that is, F is linear and $\mathcal{M}_0 = \mathcal{N} \times \{0\} \times \mathcal{P}$-monotone on all of $\mathcal{J}^2$. Each such operator is determined by the choice of a nonzero coefficient vector $J' := (a, b, E) \in \mathcal{N} \times \mathbb{R}^n \times \mathcal{P}$, that is,

$$F(r, p, A) := \langle J', J \rangle = \text{tr}(EA) + \langle b, p \rangle + ar \quad \text{for each } J := (r, p, A) \in \mathcal{J}^2. \tag{12.53}$$

Since $E \geq 0$ in $\mathcal{S}(n)$ and $a \leq 0$ in $\mathbb{R}$, one has that $(F, \mathcal{J}^2)$ is an unconstrained-case proper elliptic pair. Moreover, since F is linear and (a, b, E) is nonzero, F is topologically tame and the range $F(\mathcal{J}^2)$ is all of $\mathbb{R}$.

Theorem 12.17 (Linear equations). *Suppose that F is a proper elliptic linear operator with nonzero coefficient vector $(a, b, E) \in \mathcal{N} \times \mathbb{R}^n \times \mathcal{P}$ as defined in (12.53). Then for every $c \in \mathbb{R}$ the affine half-space*

$$\mathcal{F}_c := \{(r, p, A) \in \mathcal{J}^2 : F(r, p, A) := \operatorname{tr}(EA) + \langle b, p \rangle + ar \geq c\} \qquad (12.54)$$

is a subequation constraint set.

   For each bounded domain Ω in $\mathbb{R}^n$ one has the comparison principle

$$u \leq w \text{ on } \partial\Omega \;\Rightarrow\; u \leq w \text{ on } \Omega \qquad (12.55)$$

for $u \in \mathrm{USC}(\overline{\Omega})$ and $w \in \mathrm{LSC}(\overline{\Omega})$, which are respectively $\mathcal{F}_c$-subharmonic and $\mathcal{F}_c$-superharmonic in Ω, or equivalently, if u and w are respectively a viscosity subsolution and a viscosity supersolution to $F(u, Du, D^2u) = c$ on Ω, or, if one prefers, since the dual $\widetilde{\mathcal{F}}_c$ of $\mathcal{F}_c$ is $\mathcal{F}_{-c}$,

$$u + v \leq 0 \text{ on } \partial\Omega \;\Rightarrow\; u + v \leq 0 \text{ on } \Omega \qquad (12.56)$$

for each pair $u \in \mathcal{F}_c(\overline{\Omega})$ and $v \in \mathcal{F}_{-c}(\overline{\Omega})$.

Proof. Using definition (12.54) with $E \in \mathcal{P}$, $a \in \mathcal{N}$, and (a, b, E) nonzero, it is easy to verify that each $\mathcal{F}_c$ is a subequation. The pair $(F, \mathcal{J}^2)$ is $\mathcal{M}$-monotone for the fundamental product monotonicity subequation

$$\mathcal{M} := \mathcal{N} \times \mathcal{D}_b \times \mathcal{P} = \{(s, q, P) \in \mathcal{J}^2 : s \leq 0,\ q \in \mathcal{D}_b,\ P \in \mathcal{P}\}, \qquad (12.57)$$

where the directional cone $\mathcal{D}_b \subset \mathbb{R}^n$ is $\mathcal{D}_0 := \mathbb{R}^n$ in the gradient-free case of $b = 0$ and is the half-space

$$\mathcal{D}_b := \{q \in \mathbb{R}^n : \langle b, q \rangle \geq 0\} \qquad (12.58)$$

in the remaining case $b \neq 0$. Indeed, for each $(r, p, A) \in \mathcal{J}^2$, one has

$$F(r + s, p + q, A + P) = \operatorname{tr}(E(A + P)) + \langle b, p + q \rangle + a(r + s)$$
$$= F(r, p, A) + \operatorname{tr}(EP) + \langle b, q \rangle + as \geq F(r, p, A),$$

since $\operatorname{tr}(EP)$, $\langle b, q \rangle$, and as are all nonnegative. Hence, each $\mathcal{F}_c$ is also $\mathcal{M}$-monotone and the comparison principle (12.55) for its subharmonics and superharmonics follows from Theorem 7.6. Since F is topologically tame on $\mathcal{J}^2$, by Theorem 11.13 one will have the correspondence between $\mathcal{F}_c$-subharmonics/

superharmonics and viscosity subsolutions/supersolutions to $F = c$ since $(F, \mathcal{J}^2)$ is $\mathcal{M}$-monotone. $\qquad\square$

We now examine the linear case in more detail by showing that proper elliptic linear operators are canonical operators for the relevant half-space subequation. This will allow us to also represent certain special Hamilton–Jacobi–Bellman operators in terms of canonical operators.

Lemma 12.18 (Linear operators are canonical operators). *Let F be a proper elliptic linear operator with nonzero coefficient vector $J' = (a, b, E) \in \mathcal{N} \times \mathbb{R}^n \times \mathcal{P}$,*

$$F(J) = F(r, p, A) := \langle J', J \rangle = \mathrm{tr}(EA) + \langle b, p \rangle + ar \quad \forall J = (r, p, A) \in \mathcal{J}^2.$$
(12.59)

Then one has the following statements:

(a) *The closed linear half-space $\mathcal{F} := \{J \in \mathcal{J}^2 : F(J) := \langle J', J \rangle \geq 0\}$ with boundary orthogonal to $J' \in \mathrm{Int}\, \mathcal{F}$ is a monotonicity cone subequation which is self-dual, that is, $\widetilde{\mathcal{F}} = \mathcal{F}$.*

(b) *The maximal monotonicity cone $\mathcal{M}_{\mathcal{F}}$ of $\mathcal{F}$ as defined in Definition 4.2 is just $\mathcal{F}$ itself, that is, $\mathcal{M}_{\mathcal{F}} = \mathcal{F}$.*

(c) *Choosing any $J_0 \in \mathrm{Int}\, \mathcal{F}$, the rescaled proper elliptic linear operator with coefficient vector $J'/\langle J', J_0 \rangle$,*

$$\overline{F}(J) := \frac{1}{\langle J', J_0 \rangle} F(J) = \frac{\langle J', J \rangle}{\langle J', J_0 \rangle}, \quad J \in \mathcal{J}^2$$
(12.60)

is the canonical operator (determined by J_0) for $\mathcal{F}$ in the sense of Definition 11.16.

Proof. As noted in (12.55), with $c = 0$ the closed half-space

$$\mathcal{F} = \{J \in \mathcal{J}^2 : \langle J', J \rangle \geq 0\}$$
(12.61)

is a subequation, since it contains the minimal monotonicity set $\mathcal{M}_0 = \mathcal{N} \times \{0\} \times \mathcal{P}$. The duality claim is obvious, since $F(J) = -F(-J) = F(J)$, which completes part (a). For part (b), since $\mathcal{F}$ is a convex cone subequation, one has $\mathcal{M}_{\mathcal{F}} = \mathcal{F}$ by Proposition 4.5. Finally, for part (c), notice that the normalization of (12.60) yields the affine property

$$\overline{F}(J + tJ_0) = \frac{\langle J', J + tJ_0 \rangle}{\langle J', J_0 \rangle} = \overline{F}(J) + t \quad \text{for each } J \in \mathcal{J}^2, t \in \mathbb{R}.$$
(12.62)

The boundary of the half-space $\mathcal{F}$ is the hyperplane

$$\partial \mathcal{F} = \{J \in \mathcal{J}^2 : \overline{F}(J) = 0\}.$$
(12.63)

Since (12.62)–(12.63) hold, $\overline{F}$ is the canonical operator for $\mathcal{F}$ determined by J_0 by Proposition 11.18.   □

Remark 12.19. In the formulation of canonical operators F for a given subequation constraint set $\mathcal{F}$ which is $\mathcal{M}$-monotone, we have made an inessential normalization with respect to the affine property (12.62), which could be generalized to ask that, for some $k > 0$, one has

$$F(J + tJ_0) = F(J) + tk \quad \text{for each } J \in \mathcal{J}^2, \, t \in \mathbb{R}. \tag{12.64}$$

We have fixed this normalizing constant to be 1, but general $k > 0$ has been used in [61, Proposition 6.19] in the pure second-order case. This normalization does not affect the validity of relation (12.63) nor the other aspects of how $\partial \mathcal{F}$ decomposes $\mathcal{J}^2$, namely (see Proposition 11.18)

$$\text{Int } \mathcal{F} = \{J \in \mathcal{J}^2 : F(J) > 0\} \quad \text{and} \quad \mathcal{J}^2 \setminus \mathcal{F} = \{J \in \mathcal{J}^2 : F(J) < 0\}. \tag{12.65}$$

That is, the formula for a canonical operator F can be made to depend on both $J_0 \in \text{Int } \mathcal{M}$ and the normalizing constant k, which merely reparameterizes the distance to the boundary $\mathcal{F}$ of an $\mathcal{M}$-monotone subequation $\mathcal{F}$.

We conclude this section with a discussion of the comparison principle for a special class of Hamilton–Jacobi–Bellman equations. Admittedly, the application is perhaps a bit contrived, but it does show that certain Hamilton–Jacobi–Bellman operators are canonical for the relevant convex cone subequation. We begin by introducing the class we will treat. Let Σ be any index set and consider an arbitrary family of proper elliptic linear operators $\mathfrak{F} := \{F_\sigma : \mathcal{J}^2 \to \mathbb{R}\}_{\sigma \in \Sigma}$, where the linear operator

$$F_\sigma(J) = F_\sigma(r, p, A) := \text{tr}(E_\sigma A) + \langle b_\sigma, p \rangle + a_\sigma r = \langle J_\sigma, J \rangle \tag{12.66}$$

has nonzero coefficient vector $J_\sigma = (a_\sigma, b_\sigma, E_\sigma)$. By *proper ellipticity*,

$$a_\sigma \leq 0 \text{ in } \mathbb{R} \text{ and } E_\sigma \geq 0 \text{ in } \mathcal{S}(n), \quad \text{that is, } J_\sigma \in \mathcal{N} \times \mathbb{R}^n \times \mathcal{P} \; \forall \sigma \in \Sigma. \tag{12.67}$$

The associated linear subequations $\mathcal{F}_\sigma := \{J \in \mathcal{J}^2 : F_\sigma(J) \geq 0\}$ are the half-spaces with boundary orthogonal to J_σ.

We must also require a condition on the set of coefficient vectors in order for the intersection $\mathcal{F} := \bigcap_{\sigma \in \Sigma} \mathcal{F}_\sigma$ to be nonempty. There are several equivalent ways of formulating this condition. We start with the geometric property of being directed or pointed in the following sense.

Definition 12.20. A subset of nonzero vectors $S := \{J_\sigma\}_{\sigma \in \Sigma}$ in a finite-dimensional inner product space $(V, \langle \cdot, \cdot \rangle)$ is said to be *pointed* if, for some

$J_0 \in V \setminus \{0\}$ (called the *axis*),

$$\exists \varepsilon > 0 \text{ such that } \langle J_\sigma, J_0 \rangle \geq \varepsilon \| J_\sigma \| \quad \text{for all } \sigma \in \Sigma, \tag{12.68}$$

or equivalently (with $R := 1/\varepsilon$),

$$\exists R > 0 \text{ such that } \langle J_\sigma, J_0 \rangle > 0 \text{ and } \frac{\| J_\sigma \|}{\langle J_\sigma, J_0 \rangle} \leq R \quad \text{for all } \sigma \in \Sigma. \tag{12.69}$$

Now we examine the case of taking the infimum over such a family of proper elliptic linear operators.

Theorem 12.21 (Infimum of a family of linear operators). *Suppose that $\mathfrak{F} := \{F_\sigma\}_{\sigma \in \Sigma}$ and $\{\mathcal{F}_\sigma\}_{\sigma \in \Sigma}$ are as above.*

(a) *The intersection*

$$\mathcal{F} := \bigcap_{\sigma \in \Sigma} \mathcal{F}_\sigma \subset \mathcal{J}^2 \tag{12.70}$$

is a (convex cone) subequation if and only if

the subset $S = \{J_\sigma\}_{\sigma \in \Sigma}$ of coefficients is pointed for some axis $J_0 \neq 0$.
$$\tag{12.71}$$

(b) *Assume that the intersection $\mathcal{F}$ is a subequation with J_0 as in (12.71). Renormalize each linear operator F_σ, as in (12.60),*

$$\overline{F}_\sigma(J) := \frac{\langle J_\sigma, J \rangle}{\langle J_\sigma, J_0 \rangle} = \frac{1}{\langle J_\sigma, J_0 \rangle} F_\sigma(J), \tag{12.72}$$

to be the canonical operator for $\mathcal{F}_\sigma$ with respect to J_0. Then the infimum operator

$$F(J) := \inf_{\sigma \in \Sigma} \overline{F}_\sigma(J), \quad J \in \mathcal{J}^2 \tag{12.73}$$

is the canonical operator for the intersection subequation $\mathcal{F}$ with respect to J_0. Moreover, F is a concave function on $\mathcal{J}^2$ (and hence continuous).

Before giving the proof, a few remarks are in order.

Remark 12.22 (Geometric interpretations). Both conditions (12.68) and (12.69) for S to be pointed with axis $J_0 \neq 0$ have a geometric interpretation. First, with $\varepsilon = \| J_0 \| \cos \theta$ defining $\theta \in (0, \pi/2)$, condition (12.68) says that the angle $\sphericalangle(J_0, J_\sigma) < \theta$ for each $J_\sigma \in S$, that is, $\langle \frac{J_0}{\| J_0 \|}, \frac{J_\sigma}{\| J_\sigma \|} \rangle \geq \cos \theta$, or said differently,

$$S \text{ is contained in } C_{J_0, \theta} := \{ J \in \mathcal{J}^2 : \langle J_0, J \rangle \geq \cos \theta \| J_0 \| \, \| J \| \}, \tag{12.74}$$

where $C_{J_0,\theta}$ is called the *circular cone with axis J_0 and angle θ*. Notice that if $R \in (0, +\infty)$ is related to $\theta \in (0, 2\pi)$ by $\cos\theta = (1+R^2)^{-1/2}$, then $C_{J_0,\theta} = C_{J_0}(R)$, the cone over $\overline{B}_R(J_0/\|J_0\|)$, that is,

$$C_{J_0}(R) := \{J \in \mathcal{J}^2 : J = tJ_0 + J', \text{ where } J' \perp J_0 \text{ and } \|J'\| \le t\|J_0\|R\}.$$

Second, with the normalization $\bar{J}_\sigma := J_\sigma/\langle J_\sigma, J_0 \rangle$, so that each $\bar{J}_\sigma$ belongs to the affine hyperplane $\{J \in \mathcal{J}^2 : \langle J, J_0 \rangle = 1\}$, condition (12.69) says that

$$\{\bar{J}_\sigma\}_{\sigma \in \Sigma} \subset B_R(0), \text{ the ball of radius } R \text{ and center } 0 \text{ in } \mathcal{J}^2. \tag{12.75}$$

There are many equivalent formulations of S being pointed involving the closed convex cone hull $C(S)$ of S. For example, S is pointed if and only if $C(S)$ contains no nontrivial subspaces, that is, $C(S)$ has no *edge* (as defined in (11.71)).

Proof of Theorem 12.21. For part (a), first note that the five properties of closedness, positivity (P), negativity (N), being a cone, and being convex are all preserved under arbitrary intersections. Thus the intersection $\mathcal{F}$ is a closed convex cone satisfying (P) and (N), so that $\mathcal{F}$ is a (convex cone) subequation if and only if the topological property (T), $\mathcal{F} = \overline{\text{Int}\,\mathcal{F}}$, holds. Since $\mathcal{F}$ is closed and convex, $\mathcal{F} = \overline{\text{Int}\,\mathcal{F}}$ if and only if $\text{Int}\,\mathcal{F} \ne \emptyset$. Therefore,

$$\mathcal{F} := \bigcap_{\sigma \in \Sigma} \mathcal{F}_\sigma \text{ is a subequation } \Leftrightarrow \text{ Int}\,\mathcal{F} \ne \emptyset. \tag{12.76}$$

By the definitions, $\mathcal{F} := \bigcap_{\sigma \in \Sigma} \mathcal{F}_\sigma := \{J \in \mathcal{J}^2 : \langle J, J_\sigma \rangle \ge 0\; \forall\, \sigma \in \Sigma\} := S^\circ$, the (*convex cone*) *polar* (as defined in (11.73)) of the set of coefficient vectors $S := \{J_\sigma\}_{\sigma \in \Sigma}$. Recall that the polar of any set in an inner product space is always a closed convex cone (see Remark 11.26). The next lemma completes the proof of part (a), since by hypothesis S is pointed with axis J_0, or equivalently, $S \subset C_{J_0,\theta}$ with some $\theta \in (0, \pi/2)$, as noted in (12.74).

Lemma 12.23. *Let $S = \{J_\sigma\}_{\sigma \in \Sigma}$ be a collection of nonzero vectors in a finite-dimensional inner product space. Then*

$$J_0 \in \text{Int}\, S^\circ \;\Leftrightarrow\; S \subset C_{J_0,\theta} \text{ for some } \theta \in (0, \pi/2). \tag{12.77}$$

Proof. It is straightforward to compute the polar of a circular cone

$$C^\circ_{J_0,\theta} = C_{J_0,\theta'} \quad \text{with } \theta' = \pi/2 - \theta \in (0, \pi/2). \tag{12.78}$$

Hence, $S \subset C_{J_0,\theta}$ with $\theta \in (0, \pi/2)$ if and only if

$$C_{J_0,\theta'} \subset S^\circ = \mathcal{F} \quad \text{with } \theta' = \pi/2 - \theta \in (0, \pi/2). \tag{12.79}$$

Finally, $J_0 \in \text{Int}\,\mathcal{F} \;\Leftrightarrow\; C_{J_0,\theta'} \subset \mathcal{F}$ for some $\theta' \in (0, \pi/2)$. $\square$

For part (b), first notice that since $J_0 \in \text{Int}(\bigcap_{\sigma\in\Sigma} \mathcal{F}_\sigma) \subset \bigcap_{\sigma\in\Sigma} \text{Int}\,\mathcal{F}_\sigma$, the axis $J_0 \in \text{Int}\,\mathcal{F}_\sigma$ for each $\sigma \in \Sigma$, so that $\langle J_\sigma, J_0 \rangle > 0$. By Lemma 12.18(c), each renormalized operator $\overline{F}_\sigma$ is the canonical operator with respect to J_0 for $\mathcal{F}_\sigma$. Since the intersection $\mathcal{F}$ is a convex cone subequation, by Proposition 4.5 we have that $\mathcal{F}$ is its own maximal monotonicity cone, that is, $\mathcal{F} = \mathcal{M}_{\mathcal{F}}$ with $J_0 \in \text{Int}\,\mathcal{M}_{\mathcal{F}}$. Since $\mathcal{M}_{\mathcal{F}}$ is a monotonicity cone for the intersection $\mathcal{F}$, $\mathcal{M}_{\mathcal{F}}$ is a monotonicity cone for each $\mathcal{F}_\sigma$ and $\overline{F}_\sigma$ is the canonical operator for $\mathcal{F}_\sigma$ with respect to $J_0 \in \text{Int}\,\mathcal{M}_{\mathcal{F}}$. Since the intersection is nonempty, $F := \inf_{\sigma\in\Sigma} \overline{F}_\sigma$ is the canonical operator for the intersection $\mathcal{F}$ with respect to J_0 by Theorem 11.23(a). $\square$

Remark 12.24. Any convex cone subequation $\mathcal{F} \subset \mathcal{J}^2$ can be written as the intersection of a family of half-space subequations. As noted in the proof, the polar $\mathcal{F}^\circ$ of $\mathcal{F}$ is pointed. Choose any generating set $S := \{J_\sigma\}_{\sigma\in\Sigma}$ of nonzero vectors in $\mathcal{F}^\circ$ so that $C(S) = \mathcal{F}^\circ$. Then since $\mathcal{F}^\circ$ must be pointed by Lemma 12.23, S is a pointed set in the sense of Definition 12.20.

Armed with Theorem 12.21, we briefly discuss the comparison principle for concave Hamilton–Jacobi–Bellman operators F which are the infimum over a renormalized family of proper elliptic linear operators, whose coefficients are a pointed set $S := \{J_\sigma\}_{\sigma\in\Sigma} \subset \mathcal{N} \times \mathbb{R}^n \times \mathcal{P}$. Comparison will always hold for such operators, with possibly some restriction on the diameter of the domain Ω. The main point is contained in the following remark.

Remark 12.25. Suppose that $S := \{J_\sigma\}_{\sigma\in\Sigma} \subset \mathcal{N} \times \mathbb{R}^n \times \mathcal{P}$ is a set of nonzero coefficient vectors which is pointed with axis $J_0 \neq 0$. By Theorem 12.21, we know that the concave Hamilton–Jacobi–Bellman operator

$$F(J) := \inf_{\sigma\in\Sigma} \overline{F}_\sigma(J) := \inf_{\sigma\in\Sigma} \frac{\langle J_\sigma, J \rangle}{\langle J_\sigma, J_0 \rangle}, \quad J \in \mathcal{J}^2 \tag{12.80}$$

is the canonical operator (determined by J_0) for the monotonicity cone subequation

$$\mathcal{M}_{\mathcal{F}} = \mathcal{F} := \bigcap_{\sigma\in\Sigma} \{J \in \mathcal{J}^2 : \overline{F}_\sigma(J) \geq 0\}. \tag{12.81}$$

Hence, $(F, \mathcal{J}^2)$ is an (unconstrained-case) compatible operator–subequation pair with F topologically tame and the pair is $\mathcal{M}_{\mathcal{F}}$-monotone. Comparison on a domain Ω for viscosity subsolution/supersolution pairs of F (or for pairs of $\mathcal{M}_{\mathcal{F}} = \mathcal{F}$-subharmonics/superharmonics) reduces to the validity of the (ZMP) for the dual subequation $\widetilde{\mathcal{M}}_{\mathcal{F}}$ on Ω, which holds if one has the existence of a C^2-strict $\mathcal{M}_{\mathcal{F}}$-subharmonic ψ on Ω.

Since $\mathcal{M}_{\mathcal{F}}$ is a monotonicity cone subequation, by Theorem 5.10 it contains some element $\mathcal{M}(\gamma, \mathcal{D}, R)$ of our fundamental family with $\gamma, R \in (0, +\infty]$ and $\mathcal{D} \subseteq \mathbb{R}^n$ a directional cone. Hence, $\mathcal{M}_{\mathcal{F}}$ does indeed admit a strict approximator (with the restriction on the diameter of Ω in the case that the maximal cone $\mathcal{M}(\gamma, \mathcal{D}, R) \subset \mathcal{M}_{\mathcal{F}}$ has R finite). Hence, comparison in some form will always hold (with possible restrictions of domain diameter).

This leads to the following important question: Under what assumption on the coefficients $S := \{J_\sigma\}_{\sigma \in \Sigma}$ will we have a given inclusion

$$\mathcal{M} \subset \mathcal{M}_{\mathcal{F}} \text{ for a given monotonicity cone subequation } \mathcal{M}? \tag{12.82}$$

The needed inclusion (12.82) is equivalent to the reverse inclusion for the polars

$$\mathcal{M}_{\mathcal{F}}^\circ \subset \mathcal{M}^\circ, \tag{12.83}$$

and recalling that $S^\circ = \mathcal{M}_{\mathcal{F}}$, a necessary and sufficient condition on S in order to have (12.82) is

$$S \subset \mathcal{M}^\circ. \tag{12.84}$$

As noted before, since $\mathcal{M}_0 := \mathcal{N} \times \{0\} \times \mathcal{P} \subset \mathcal{M}$ for every monotonicity cone subequation $\mathcal{M}$, one must have the set S of coefficient vectors contained in

$$\mathcal{M}^\circ \subset \mathcal{M}_0^\circ = \mathcal{N} \times \mathbb{R}^n \times \mathcal{P},$$

that is, a set S of proper elliptic coefficient vectors in $\mathcal{J}^2$.

Combining Theorem 12.21 with the considerations of Remark 12.25 yields the following result. The reader might wish to consult Definition 5.2 and Remark 5.9 to review the family of monotonicity cones, as well as Theorem 6.3 on the validity of the (ZMP) for the duals of our family of monotonicity cone subequations $\mathcal{M}$.

Theorem 12.26 (Comparison for the inf of a pointed family of linear operators). *Suppose that $\{J_\sigma\}_{\sigma \in \Sigma}$ is a pointed set (with axis $J_0 \neq 0$) of nonzero vectors in $\mathcal{N} \times \mathbb{R}^n \times \mathcal{P} \subset \mathcal{J}^2$. Consider the associated (normalized) proper elliptic linear operators*

$$\overline{F}_\sigma(J) := \frac{\langle J_\sigma, J \rangle}{\langle J_\sigma, J_0 \rangle} = \frac{1}{\langle J_\sigma, J_0 \rangle} F_\sigma(J) \tag{12.85}$$

and half-space subequations

$$\mathcal{F}_\sigma := \{J \in \mathcal{J}^2 : \overline{F}_\sigma(J) \geq 0\},$$

whose intersection $\mathcal{M}_{\mathcal{F}} = \mathcal{F} := \bigcap_{\sigma \in \Sigma} \mathcal{F}_\sigma$ is a convex cone subequation for which

$$F(J) := \inf_{\sigma \in \Sigma} \overline{F}_\sigma(J)$$

is the canonical operator (determined by J_0) for $\mathcal{M}_{\mathcal{F}} = \mathcal{F}$.

(a) *Suppose that the coefficient vectors satisfy $S \subset \mathcal{M}^\circ$ with $\mathcal{M}$ being one of the monotonicity cone subequations $\mathcal{M}(\gamma)$, $\mathcal{M}(\mathcal{P})$, $\mathcal{M}(\mathcal{D})$, $\mathcal{M}(\gamma, \mathcal{D})$, $\mathcal{M}(\mathcal{D}, \mathcal{P})$, or $\mathcal{M}(\gamma, \mathcal{D}, \mathcal{P})$ (the case $R = +\infty$ of Theorem 6.3). Then for every $c \in \mathbb{R}$ one has the comparison principle*

$$u \leq w \text{ on } \partial\Omega \ \Rightarrow \ u \leq w \text{ on } \Omega \tag{12.86}$$

for $u \in \mathrm{USC}(\overline{\Omega})$ and $w \in \mathrm{LSC}(\overline{\Omega})$, which are a viscosity subsolution/super-solution pair for the Hamilton–Jacobi–Bellman equation $F(u, Du, D^2u) = c$. In this case, Ω is an arbitrary bounded domain.

(b) *If, instead, $S \subset \mathcal{M}^\circ$ with $\mathcal{M}$ being one of the monotonicity cone subequations $\mathcal{M}(R)$, $\mathcal{M}(\gamma, R)$, $\mathcal{M}(\mathcal{D}, R)$, or $\mathcal{M}(\gamma, \mathcal{D}, R)$ with R finite (the case R finite of Theorem 6.3), then the comparison principle for F holds on domains contained in a translate of the truncated cone $\mathcal{D} \cap B_R(0)$, which is a ball of radius R in the case $\mathcal{D} = \mathbb{R}^n$.*

In order to implement Theorem 12.26, which requires condition (12.84) on the coefficients, that is, $S \subset \mathcal{M}^\circ$, we list some of the polars of our family of monotonicity cone subequations $\mathcal{M}$. The proof is left to the reader.

Proposition 12.27 (Polars of some monotonicity cone subequations). *One has the following polar formulas:*

(a) *For $\mathcal{M}(\mathcal{P}) := \{(r, p, A) \in \mathcal{J}^2 : A \in \mathcal{P}\}$, one has*

$$\mathcal{M}(\mathcal{P})^\circ = \{0\} \times \{0\} \times \mathcal{P}. \tag{12.87}$$

(b) *For $\mathcal{M}(R) := \{(r, p, A) \in \mathcal{J}^2 : A \geq \frac{|p|}{R} I\}$ with $R \in (0, +\infty)$, one has*

$$\mathcal{M}(R)^\circ = \{(s, q, B) \in \mathcal{J}^2 : s = 0, \ B \geq 0, \ \text{and } \mathrm{tr}\, B \geq R|q|\}. \tag{12.88}$$

(c) *For $\mathcal{M}(\gamma) := \{(r, p, A) \in \mathcal{J}^2 : r \leq -\gamma|p|\}$ with $\gamma \in [0, +\infty)$, one has*

$$\mathcal{M}(\gamma)^\circ = M'(1/\gamma) \times \{0\}$$
$$= \{(s, q, B) \in \mathcal{J}^2 : B = 0, \ s \leq -\tfrac{1}{\gamma}|q|\}, \tag{12.89}$$

which includes $\mathcal{M}(\mathcal{N})^\circ = \mathcal{N} \times \{0\} \times \{0\}$ in the case $\gamma = 0$.

(d) *For $\mathcal{M}(\gamma, R) := \{(r, p, A) \in \mathcal{J}^2 : r \leq -\gamma|p| \text{ and } A \geq \frac{|p|}{R}I\}$ with $\gamma \in [0, +\infty)$ and $R \in (0, +\infty]$, one has*

$$\mathcal{M}(\gamma, R)^\circ = \{(s, q, B) \in \mathcal{J}^2 : B \geq 0 \text{ and } \operatorname{tr} B \geq R(|q| + \gamma s)\}, \qquad (12.90)$$

which includes $\mathcal{M}(\mathcal{N}, R)^\circ = \{(s, q, B) \in \mathcal{J}^2 : s \leq 0, \ B \geq 0, \text{ and } \operatorname{tr} B \geq R|q|\}$ in the case $\gamma = 0$.

(e) *For $\mathcal{M}(\gamma, \mathcal{P}) := \{(r, p, A) \in \mathcal{J}^2 : r \leq -\gamma|p| \text{ and } A \geq 0\}$ with $\gamma \in [0, +\infty)$, one has*

$$\mathcal{M}(\gamma, \mathcal{P})^\circ = \{(s, q, B) \in \mathcal{J}^2 : B \geq 0 \text{ and } r \leq -\frac{1}{\gamma}|q|\}. \qquad (12.91)$$

(f) *For $\mathcal{M}(\mathcal{N}, \mathcal{P}) := \mathcal{N} \times \mathbb{R}^n \times \mathcal{P} = \{(r, p, A) \in \mathcal{J}^2 : r \leq 0 \text{ and } A \geq 0\}$, one has*

$$\begin{aligned} \mathcal{M}(\mathcal{N}, \mathcal{P})^\circ &= \mathcal{N} \times \{0\} \times \mathcal{P} \\ &= \{(s, q, B) \in \mathcal{J}^2 : s \leq 0 \text{ and } B \geq 0\}. \end{aligned} \qquad (12.92)$$

(g) *For $\mathcal{M}(\mathcal{N}, \mathcal{D}, \mathcal{P}) := \mathcal{N} \times \mathcal{D} \times \mathcal{P}$ with $\mathcal{D} \subsetneq \mathbb{R}^n$, one has*

$$\mathcal{M}(\mathcal{N}, \mathcal{D}, \mathcal{P})^\circ = \mathcal{N} \times \mathcal{D}^\circ \times \mathcal{P}. \qquad (12.93)$$

Finally, for the intersection of any of the cones $\mathcal{M}$ in cases (a)–(f) with $\mathcal{M}(\mathcal{D})$ (with $\mathcal{D} \subsetneq \mathbb{R}^n$), the polar $(\mathcal{M} \cap \mathcal{M}(\mathcal{D}))^\circ$ can be expressed as $\overline{(\mathcal{M}^\circ + \mathcal{M}(\mathcal{D})^\circ)}$, but an explicit description is more complicated and is left to the interested reader.

We now present two representative examples of pointed families $S = \{J_\sigma\}_{\sigma \in \Sigma}$ which give rise to comparison for the associated Hamilton–Jacobi–Bellman operators, with and without restrictions on the size of the domain Ω. First, we give a simple example where there is an a priori restriction on the size of the domain.

Example 12.28. For $R \in (0, +\infty)$, consider the convex cone subequation defined in (5.2) and (9.1),

$$\mathcal{M}(R) = \mathcal{F}_{1,R}^- := \{(r, p, A) \in \mathcal{J}^2 : \lambda_1(A) - \frac{|p|}{R} \geq 0\}, \qquad (12.94)$$

where $\lambda_1(A)$ is the smallest eigenvalue of $A \in \mathcal{S}(n)$. Consider the index set

$$\Sigma := S^{n-1} \times S^{n-1} = \{\sigma = (\xi, \eta) \in \mathbb{R}^n \times \mathbb{R}^n : |\xi| = 1 = |\eta|\}.$$

Since for all $A \in \mathcal{S}(n)$ and $p \in \mathbb{R}^n$,

$$\lambda_1(A) = \inf_{\xi \in S^{n-1}} \operatorname{tr}((\xi \otimes \xi)A) \quad \text{and} \quad |p| = -\inf_{\eta \in S^{n-1}} \langle \eta, p \rangle,$$

where $\mathrm{tr}((\xi \otimes \xi)A) = \langle A\xi, \xi \rangle = \langle A, P_\xi \rangle$, one has that

$$\mathcal{F}_{1,R}^{-} = \bigcap_{\sigma \in \Sigma} \mathcal{F}_\sigma,$$

where

$$\mathcal{F}_\sigma := \{ (r, p, A) \in \mathcal{J}^2 : \langle J_\sigma, J \rangle \ge 0 \} \quad \text{with } J_\sigma := \left(0, \frac{\eta}{R}, \xi \otimes \xi \right).$$

One can easily check that the set of coefficient vectors $S := \{ J_\sigma \}_{\sigma \in \Sigma}$ is pointed with axis $J_0 = (0, 0, I)$. Furthermore, since $\langle J_\sigma, J_0 \rangle = 1$ for all $\sigma \in \Sigma$, the infimum operator defined in (12.73) is given by

$$F(J) = \inf_{\sigma \in \Sigma} \frac{\langle J_\sigma, J \rangle}{\langle J_\sigma, J_0 \rangle} = \lambda_1(A) - \frac{|p|}{R}.$$

Finally, we know that the polar S° of the set S of coefficient vectors is the maximal monotonicity cone $\mathcal{M}_{\mathcal{F}_{1,R}^{+}}$ of $\mathcal{F}_{1,R}^{+}$, which was shown in Proposition 9.2(a) to satisfy $\mathcal{M}_{\mathcal{F}_{1,R}^{+}} = \mathcal{M}(R)$ and hence $S^\circ = C(S)^\circ = \mathcal{M}(R)$, where $C(S)$ is the closed convex hull of S. By the bipolar theorem, one has $S \subset C(S) = \mathcal{M}(R)^\circ$. Hence, by Theorem 12.21(b), the comparison principle for the equation $F(u, Du, D^2u) = c$ holds on domains Ω contained in a ball of radius R.

We now turn to a class of Hamilton–Jacobi–Bellman equations for which comparison holds in arbitrary bounded domains. These examples are motivated by a question of optimal control. A good general reference for this subject is the monograph of Bardi–Capuzzo Dolcetta [3].

Example 12.29. One important problem in optimal control concerns an agent who seeks to minimize an infinite-horizon discounted cost functional by acting on its drift and volatility parameters. We consider infima of linear operators of the form

$$F_\sigma(J) = F_\sigma(r, p, A) = \mathrm{tr}(E_\sigma A) + \langle b_\sigma, p \rangle + cr = \langle J_\sigma, J \rangle, \quad \sigma \in \Sigma, \qquad (12.95)$$

where $\delta := -c > 0$ is the *discount factor*, b_σ is the *drift term*, and E_σ is the *(squared) volatility*. In this example, E_σ is allowed to vary in bounded sets, that is, for some $m > 0$,

$$\|E_\sigma\| \le m \quad \text{for all } \sigma \in \Sigma. \qquad (12.96)$$

The set of drifts $S_d := \{ b_\sigma \}_{\sigma \in \Sigma}$ will be taken to be pointed with axis $b_0 \in \mathbb{R}^n \setminus \{0\}$, that is,

$$\exists \, \varepsilon' > 0 \text{ such that } \langle b_0, b_\sigma \rangle \ge \varepsilon' |b_\sigma| \quad \text{for all } \sigma \in \Sigma, \qquad (12.97)$$

which means that all possible drifts share a "preferred" direction b_0. We will denote by

$$\mathcal{D} := C(S_d) \tag{12.98}$$

the closed convex hull of S_d, whose polar $\mathcal{D}^\circ$ agrees with the polar S_d° of S_d.

The set of coefficient vectors $S = \{J_\sigma\}_{\sigma\in\Sigma}$ is pointed with axis $J_0 = (-1, b_0, I)$ in the sense of (12.68), that is, there exists $\varepsilon > 0$ such that

$$\langle J_\sigma, J_0 \rangle \geq \varepsilon \|J_\sigma\| \quad \text{for all } \sigma \in \Sigma. \tag{12.99}$$

In fact, using (12.97), $c \leq 0$, and $E_\sigma \geq 0$ is $\mathcal{S}(n)$, one has

$$\langle J_\sigma, J_0 \rangle = |c| + \langle b_\sigma, b_0 \rangle + \operatorname{tr} E_\sigma \geq |c| + \varepsilon' |b_\sigma| + C_n \|E_\sigma\| > 0 \tag{12.100}$$

for some dimensional constant $C_n > 0$. Since

$$\varepsilon^2 \|J_\sigma\|^2 = \varepsilon^2 (c^2 + |b_\sigma|^2 + \|E_m\|^2), \tag{12.101}$$

squaring the inequality in (12.100) and comparing with (12.101) shows that (12.99) holds for $\varepsilon \leq \min\{1, \varepsilon', C_n\}$.

Since the set of coefficient vectors $S = \{J_\sigma\}_{\sigma\in\Sigma}$ is pointed, the intersection

$$\mathcal{F} = \bigcap_{\sigma\in\Sigma} \{(r, p, A) \in \mathcal{J}^2 : \langle J_\sigma, J \rangle \geq 0\}$$

is a convex cone subequation with maximal monotonicity cone $\mathcal{M}_\mathcal{F} = \mathcal{F}$ and the infimum operator defined by

$$F(J) = \inf_{\sigma\in\Sigma} \frac{\langle J_\sigma, J \rangle}{\langle J_\sigma, J_0 \rangle}$$

is the canonical operator for $\mathcal{F}$ by Theorem 12.21. Finally, since $S_d \subset C(S_d) := \mathcal{D}$, one has

$$S \subset C(S) \subset \mathcal{N} \times \mathcal{D} \times \mathcal{P} = (\mathcal{N} \times \mathcal{D}^\circ \times \mathcal{P})^\circ = \mathcal{M}^\circ$$

for the monotonicity cone subequation $\mathcal{M} := \mathcal{M}(\mathcal{N}, \mathcal{D}, \mathcal{P}) = \mathcal{M}(\gamma, \mathcal{D}, \mathcal{P})$ with $\gamma = 0$. Hence, one has the comparison principle for the Hamilton–Jacobi–Bellman equation $F(u, Du, D^2 u) = c$ for each $c \in \mathbb{R}$ by Theorem 12.26.

We conclude this section with the following observations.

Remark 12.30 (More general convex subequations as intersections). Choosing a pointed family $\{F_\sigma\}_{\sigma\in\Sigma}$ of linear operators as above, and then choosing constants $\{c_\sigma\}_{\sigma\in\Sigma}$, the affine half-spaces $\mathcal{F}_\sigma := \{J \in \mathcal{J}^2 : F_\sigma(J) \geq c_\sigma\}$ have intersection $\mathcal{F} := \bigcap_{\sigma\in\Sigma} \mathcal{F}_\sigma$, which is a convex subequation, and all convex subequations arise in this manner. Moreover, Theorem 12.26 carries to these more general convex (but not cone) subequations $\mathcal{F}$. The details are left to the reader.

Remark 12.31 (Duality for pointed families of linear operators). Linear operators F and their half-space subequations $\mathcal{F} := \{F(J) \geq 0\}$ are self-dual, while the dual of $\mathcal{F}_c := \{F(J) \geq c\}$ is $\mathcal{F}_{-c} := \{F(J) \geq -c\}$. Therefore, given a family of such operators $\mathfrak{F} := \{F_\sigma\}_{\sigma \in \Sigma}$ as in Theorem 12.21, the list of four subequations in the general theorem (Theorem 11.23, on duality, intersections, and unions) reduces to two,

$$\mathcal{F} := \bigcap_{\sigma \in \Sigma} \mathcal{F}_\sigma \quad \text{and} \quad \mathcal{E} := \overline{\bigcup_{\sigma \in \Sigma} \mathcal{F}_\sigma}. \tag{12.102}$$

If the operators F_σ are normalized to be canonical with respect to J_0 as in (12.85), then

$$F := \inf_{\sigma \in \Sigma} F_\sigma \quad \text{and} \quad E := \sup_{\sigma \in \Sigma} F_\sigma \tag{12.103}$$

are the corresponding canonical operators for $\mathcal{F}$ and $\mathcal{E}$.

A comparison principle analogous to Theorem 12.26 for F, $\mathcal{F}$ holds for E, $\mathcal{E}$. This is left to the interested reader. Also, as in Remark 12.30, constants $\{c_\sigma\}_{\sigma \in \Sigma}$ can be employed, yielding comparison for appropriate unions $\mathcal{E}$ of affine half-spaces and a supremum operator E. This is, of course, no surprise, as $\mathcal{E}$ and E are dual to $\mathcal{F}$ and F.

12.5 PROPER ELLIPTIC OPERATORS WITH DIRECTIONALITY IN THE GRADIENT

Next we consider proper elliptic operators whose monotonicity also includes directionality in the gradient with respect to a directional cone $\mathcal{D} \subsetneq \mathbb{R}^n$ when $\mathcal{D}$ is a proper subset. More precisely, we will consider proper elliptic pairs $(F, \mathcal{F})$ when the subequation constraint set $\mathcal{F}$ satisfies the directionality condition (D),

$$(r, p, A) \in \mathcal{F} \;\Rightarrow\; F(r, p+q, A) \in \mathcal{F} \text{ for each } q \in \mathcal{D}, \tag{12.104}$$

and the operator F has the corresponding monotonicity in the gradient variable,

$$F(r, p, A) \leq F(r, p+q, A) \quad \text{for each } (r, p, A) \in \mathcal{F}, \, q \in \mathcal{D}. \tag{12.105}$$

For simplicity, we will focus on the fundamental product monotonicity cone given by $\mathcal{M} := \mathcal{N} \times \mathcal{D} \times \mathcal{P}$ and prove comparison on arbitrary bounded domains in both constrained and unconstrained cases with $\mathcal{M}$-monotonicity. When the operator F needs to be constrained in order to be proper elliptic, there is a dichotomy similar to what we have seen in the gradient-free case. More precisely, operators such as

$$F(r, p, A) = -rp_n \det(A), \quad \text{with } \mathcal{D} = \{p = (p', p_n) \in \mathbb{R}^n : p_n \geq 0\}, \quad (12.106)$$

are compatible with $\mathcal{F} = \mathcal{M} = \mathcal{N} \times \mathcal{D} \times \mathcal{P}$, while operators such as

$$F(r, p, A) = -r \det(A) - p_n \quad (12.107)$$

do not satisfy compatibility with respect to the constraint $\mathcal{F} = \mathcal{M} = \mathcal{N} \times \mathcal{D} \times \mathcal{P}$ (or any other constraint for that matter). We leave it to the reader to verify this last claim, making use of the discussion in Remark 12.7 on the gradient-free case. Although example (12.107) behaves badly, in Theorem 12.36 below we describe a general class of operators $F(r, p, A) := G(r, p', A) - p_n$ which are good unconstrained-case examples (where comparison holds).

Examples that we will treat include equations of Monge–Ampère type that arise in optimal transport (see Example 12.34 below). All of the examples we treat in this section (with $\mathcal{D} \subsetneq \mathbb{R}^n$) are "very weakly" parabolic. This will lead to the treatment of genuinely parabolic equations in the next section (see Example 12.35 below).

We begin with a general class of operators that includes (12.106), before discussing good versions of (12.107). Recall that $\mathcal{D} \subsetneq \mathbb{R}^n$ is a *directional cone* if it is a closed convex cone with vertex at the origin (see Definition 2.2).

Theorem 12.32. *Consider the operator defined by*

$$F(r, p, A) = d(p)G(r, A). \quad (12.108)$$

Suppose that $(G, \mathcal{G})$ is a (constrained-case) compatible proper elliptic gradient-free pair in the sense of Definition 12.1, with the normalization

$$\inf_{\mathcal{G}} G = 0, \quad \text{so that } \partial\mathcal{G} = \{(r, A) \in \mathcal{G} : G(r, A) = 0\}. \quad (12.109)$$

Given $\mathcal{D} \subsetneq \mathbb{R}^n$ a directional cone (which is then a $\mathcal{D}$-monotone pure first-order subequation) and given a continuous function $d : \mathcal{D} \to \mathbb{R}$, suppose that d is $\mathcal{D}$-monotone,

$$d(p + q) \geq d(p) \quad \text{for each } p, q \in \mathcal{D}, \quad (12.110)$$

and suppose that $(d, \mathcal{D})$ is a compatible pair, that is,

$$d(p) \geq 0 \text{ for all } p \in \mathcal{D} \quad \text{and} \quad d(p) = 0 \text{ if and only if } p \in \partial\mathcal{D}. \quad (12.111)$$

Then $(F, \mathcal{F})$ is a compatible proper elliptic pair for the subequation $\mathcal{F}$ defined by

$$\mathcal{F} := \{(r, p, A) \in \mathcal{J}^2 : p \in \mathcal{D}, \ (r, A) \in \mathcal{G}, \text{ and } F(r, p, A) \geq 0\}, \quad (12.112)$$

and the pair is $\mathcal{M}$-monotone for the monotonicity cone subequation

$$\mathcal{M} = \mathcal{N} \times \mathcal{D} \times \mathcal{P} := \{(s, q, P) : s \leq 0, \ p \in \mathcal{D}, \ P \in \mathcal{P}\}. \tag{12.113}$$

Consequently, for every $c \in F(\mathcal{F})$ and for every bounded domain $\Omega \subset \mathbb{R}^n$, one has the comparison principle

$$u \leq w \ on \ \partial\Omega \ \Rightarrow \ u \leq w \ on \ \Omega \tag{12.114}$$

for $u \in \mathrm{USC}(\overline{\Omega})$ and $w \in \mathrm{LSC}(\overline{\Omega})$, which are respectively $\mathcal{F}_c$-subharmonic and $\mathcal{F}_c$-superharmonic in Ω, where $\mathcal{F}_c := \{(r, p, A) \in \mathcal{F} : F(r, p, A) \geq c\}$.

If one also requires that F is topologically tame on $\mathcal{F}$, then the comparison principle (12.114) equivalently holds if u and w are respectively an $\mathcal{F}$-admissible subsolution and an $\mathcal{F}$-admissible supersolution to $F(u, Du, D^2u) = c$ on Ω.

Proof. Being proper elliptic, the pair $(G, \mathcal{G})$ is $\mathcal{Q}$-monotone, which together with the sign conditions $d \geq 0$ on $\mathcal{D}$ and $G \geq 0$ on $\mathcal{G}$ easily leads to $(F, \mathcal{F})$ being a proper elliptic pair. Indeed, if $(r, p, A) \in \mathcal{F}$ then $p \in \mathcal{D}$, $(r, A) \in \mathcal{G}$, and $d(p)G(r, A) \geq 0$, so that, for each $s \leq 0$ and $P \in \mathcal{P}$, one has $(r + s, A + P) \in \mathcal{G}$ and

$$F(r + s, p, A + P) = d(p)G(r + s, A + P) \geq d(p)G(r, A) = F(r, p, A) \geq 0.$$

The pair is compatible with

$$\inf_{\mathcal{F}} F = 0 \quad \text{and} \quad \partial\mathcal{F} = \{(r, p, A) \in \mathcal{F} : F(r, p, A) = 0\},$$

where one uses (12.109) and (12.111).

The pair $(F, \mathcal{F})$ is $\mathcal{M}$-monotone for the monotonicity cone subequation (12.113). Indeed, for each $(r, p, A) \in \mathcal{F}$ and each $(s, q, P) \in \mathcal{M}$, one has

$$F(r + s, p + q, A + P) = d(p + q)G(r + s, A + P) \geq d(p)G(r, A) = F(r, p, A) \tag{12.115}$$

by the proper ellipticity of G on $\mathcal{G}$ and the directionality condition (12.110). Hence, by Lemma 11.3, for each upper level set, $\mathcal{F}_c$ in (12.115) is $\mathcal{M}$-monotone.

Finally, if the operator F is topologically tame (see Definition 11.9), that is, for every admissible level $c \in F(\mathcal{F})$, the interior of

$$\mathcal{F}(c) := \{(r, p, A) \in \mathcal{F} : F(r, p, A) = c\}$$

is empty, then one has the correspondence principle of Theorem 11.13 and hence the comparison principle for $\mathcal{F}$-admissible subsolutions, supersolutions u, w. $\square$

Now we give some explicit examples where this Theorem 12.32 applies, and we make a few observations.

Example 12.33. Start with one of the gradient-free pairs $(G, \mathcal{G})$,

$$G(r, A) = h(r)\mathfrak{g}(A) \quad \text{and} \quad \mathcal{G} = \mathbb{N} \times \overline{\Gamma} \subset \mathbb{R} \times \mathcal{S}(n) \tag{12.116}$$

of Example 12.5, where $\mathfrak{g}$ is a Gårding–Dirichlet polynomial and $h \in C((-\infty, 0])$ is nonnegative, nondecreasing, and satisfies $h(r) = 0$ if and only if $r = 0$. The prototype is $G(r, A) := -r \det(A)$ as in (12.106). As for the pair $(d, \mathcal{D})$, we mention

$$d(p) = p_n \quad \text{and} \quad \mathcal{D} = \{(p', p_n) \in \mathbb{R}^n : p_n \geq 0\} \quad \text{(a half-space),} \tag{12.117}$$

$$d(p) = \prod_{j=1}^{n} p_j \quad \text{and} \quad \mathcal{D} = \{(p_1, \ldots, p_n) \in \mathbb{R}^n : p_j \geq 0 \text{ for each } j = 1, \ldots, n\}, \tag{12.118}$$

and more generally, for some $k \in \{1, \ldots, n\}$,

$$d(p) = \prod_{j=1}^{k} p_j \quad \text{and} \quad \mathcal{D} = \{(p_1, \ldots, p_n) \in \mathbb{R}^n : p_j \geq 0 \text{ for each } j = 1, \ldots, k\}. \tag{12.119}$$

Now set $F(r, p, A) := d(p)G(r, A)$. In all of these examples, F is topologically tame since F is real analytic.

An interesting special case comes from a very special form of *optimal transport*, an important subject which has received much recent attention. Excellent general references include the monograph of Villani [89] and the survey paper of De Fillipis–Figalli [32].

Example 12.34 (Optimal transport with uniform source density). A partial differential equation of the form

$$d(Du) \det(D^2 u) = c, \quad c \geq 0 \tag{12.120}$$

arises in the theory of optimal transport, under some restrictive assumptions. In general, one has a function $f = f(x)$ in place of the constant c, where f represents the mass density in the source configuration and d represents the mass density of the target configuration (with the mass balance $\|f\|_{L^1} = \|d\|_{L^1}$). One seeks to transport the mass with density f onto the mass with density d at minimal transportation cost (which is quadratic with respect to transport distance). The solution of this minimization problem is given by the gradient of a convex function u, which turns out to be a generalized solution of equation (12.120). In the special case of uniform source density $f \equiv c$ and with target density d having some directionality, comparison principles can be obtained as a special case of Example 12.33 with $h(r) :\equiv 1$ and $\mathfrak{g}(A) := \det A$.

Example 12.35. In the case where the gradient factor d is defined by (12.117) and $G(r, A)$ depends only on $A' \in \mathcal{S}(n-1)$ (second-order derivatives only in the *spatial variables* $x' \in \mathbb{R}^{n-1}$), one has a fully nonlinear *parabolic* equation of the kind considered by Krylov in his extension of Alexandroff's methods to parabolic equations in [75]. Such genuinely parabolic situations will be discussed in the next section. For the example here, it can be treated by Theorem 12.38 below.

We now treat the unconstrained case, which takes into account the difficulty posed by operators such as the one defined in (12.107). The operators are proper elliptic with strict monotonicity in a gradient variable.

Theorem 12.36. *Suppose that $F \colon \mathcal{J}^2 \to \mathbb{R}$ is continuous and of the form*

$$F(r, p, A) := G(r, p', A) - p_n, \tag{12.121}$$

where $G \colon \mathbb{R} \times \mathbb{R}^{n-1} \times \mathcal{S}(n) \to \mathbb{R}$ is continuous and satisfies the two conditions of minimal $(\mathcal{N} \times \{0\} \times \mathcal{P})$-monotonicity,

$$G(r, p', A) \leq G(r+s, p', A+P) \quad \text{for each } s \leq 0, \ P \in \mathcal{P}, \tag{12.122}$$

and directional monotonicity for some $\beta > 0$ fixed,

$$G(r, p'+q', A) - G(r, p', A) \geq -\beta |q'| \quad \text{for each } q' \in \mathbb{R}^{n-1}, \tag{12.123}$$

where the directional cone $\mathcal{D} \subsetneq \mathbb{R}^n = \mathbb{R}^{n-1} \times \mathbb{R}$ is the circular cone defined by

$$\mathcal{D} := \{q = (q', q_n) \in \mathbb{R}^n : -q_n \geq \beta |q'|\}. \tag{12.124}$$

Then $(F, \mathcal{J}^2)$ is an unconstrained case of a proper elliptic pair which is $\mathcal{M}$-monotone for the monotonicity cone

$$\mathcal{M} := \{(s, q, P) \in \mathcal{J}^2 : s \leq 0, \ q \in \mathcal{D}, \ \text{and } P \in \mathcal{P}\} = \mathcal{N} \times \mathcal{D} \times \mathcal{P}. \tag{12.125}$$

Consequently, for every $c \in F(\mathcal{J}^2)$ and for every bounded domain $\Omega \subset \mathbb{R}^n$, one has the comparison principle

$$u \leq w \text{ on } \partial\Omega \ \Rightarrow \ u \leq w \text{ on } \Omega \tag{12.126}$$

for $u \in \mathrm{USC}(\overline{\Omega})$ and $w \in \mathrm{LSC}(\overline{\Omega})$, which are respectively $\mathcal{F}_c$-subharmonic and $\mathcal{F}_c$-superharmonic in Ω with $\mathcal{F}_c := \{(r, p, A) \in \mathcal{J}^2 : F(r, p, A) \geq c\}$.

If one also requires that F is topologically tame on $\mathcal{F}$, then the comparison principle (12.126) equivalently holds if u and w are respectively a viscosity subsolution and a viscosity supersolution to $F(u, Du, D^2u) = c$ on Ω.

Proof. One follows the same argument as the proofs of Theorems 12.8 and 12.11. It is sufficient to observe that if $(r, p, A) \in \mathcal{F}$ and $(s, q, P) \in \mathcal{M}$, then by (12.122) and (12.123) one has

$$\begin{aligned} F(r+s, p+q, A+P) &= G(r+s, p'+q', A+P) - (p_n + q_n) \\ &\geq G(r, p'+q', A) - (p_n + q_n) \\ &\geq G(r, p', A) - p_n - q_n - \beta |q'| \geq 0, \end{aligned}$$

since $(r, p, A) \in \mathcal{F}$ and $q = (q', q_n) \in \mathcal{D}$. Therefore, $(F, \mathcal{J}^2)$ is proper elliptic and $\mathcal{M}$-monotone. By Lemma 11.3, each $\mathcal{F}_c$ is $\mathcal{M}$-monotone and the comparison principle for each $c \in F(\mathcal{J}^2)$ follows from Theorem 7.6, since $\mathcal{M}$ is a basic product monotonicity cone. Finally, if F is topologically tame, the correspondence principle of Theorem 11.13 applies to yield comparison for (standard) viscosity subsolutions and supersolutions of the equation $F(u, Du, D^2u) = c$ for each $c \in F(\mathcal{J}^2)$. $\qquad\square$

This section highlights one interesting feature of the approach to comparison principles by monotonicity cones and duality. Namely, "very weakly" parabolic equations are included naturally in the general theory. We will continue to develop this idea in the next section, which includes genuinely parabolic equations with a parabolic form of the comparison principle.

12.6   PARABOLIC OPERATORS

If F is genuinely parabolic in the sense that G as in (12.108) or (12.121) depends only on derivatives with respect to the *spatial variables* $x' \in \mathbb{R}^{n-1}$, we establish a genuinely parabolic version of the comparison principle involving the parabolic boundary. First, we give some notation. In $\mathbb{R}^n = \mathbb{R}^{n-1} \times \mathbb{R}$, in which both points of Ω and the gradient variables live, we will drop the primes and replace (x', x_n) and (p', p_n) by (x, t) and (p, τ) respectively. In the matrix variable $\mathcal{S}(n)$, we will use the primes for elements and subsets of $\mathcal{S}(n-1)$ as follows. For $A \in \mathcal{S}(n)$ we will denote by A' the $(n-1) \times (n-1)$ minor in the x-variables, that is,

$$A = \begin{pmatrix} A' & * \\ * & * \end{pmatrix}.$$

We will also need the matrix sets

$$\mathcal{P}' := \{ P' \in \mathcal{S}(n-1) : P' \geq 0 \} \tag{12.127}$$

and

$$\mathcal{P}^* := \{ A \in \mathcal{S}(n) : A' \in \mathcal{P}' \}. \tag{12.128}$$

Finally, one defines the *parabolic boundary of* $\Omega \subset \mathbb{R}^n$ by

$$\partial^- \Omega := \{(x,t) \in \partial\Omega : t < T\}, \quad \text{where } T := \sup\{t \in \mathbb{R} : (x,t) \in \Omega\}. \quad (12.129)$$

We begin with the parabolic version of the unconstrained case of Theorem 12.36.

Theorem 12.37. *Suppose that* $F \colon \mathcal{J}^2 \to \mathbb{R}$ *is continuous and of the form*

$$F(r, (p, \tau), A) := G(r, p, A') - \tau, \quad (12.130)$$

where $G \colon \mathbb{R} \times \mathbb{R}^{n-1} \times \mathcal{S}(n-1) \to \mathbb{R}$ *is continuous and satisfies for each* $(r, p, A') \in \mathbb{R} \times \mathbb{R}^{n-1} \times \mathcal{S}(n-1)$ *the two conditions, first of* $\mathcal{N} \times \{0\} \times \mathcal{P}'$-*monotonicity,*

$$G(r, p, A') \leq G(r + s, p, A' + P') \quad \text{for each } s \leq 0, \ P' \in \mathcal{P}, \quad (12.131)$$

and second for some $\gamma > 0$ *fixed,*

$$G(r, p+q, A') - G(r, p, A') \geq -\gamma |q| \quad \text{for each } q \in \mathbb{R}^{n-1}, \quad (12.132)$$

where $\mathcal{D} \subset \mathbb{R}^n$ *is the circular directional cone of* (12.124), *that is, with the current notation,*

$$\mathcal{D} := \{(q, \sigma) \in \mathbb{R}^{n-1} \times \mathbb{R} : -\sigma \geq \gamma |q|\}. \quad (12.133)$$

Then $(F, \mathcal{J}^2)$ *is an unconstrained proper elliptic pair which is* $\mathcal{M}^*$-*monotone for the monotonicity cone*

$$\mathcal{M}^* = \mathcal{N} \times \mathcal{D} \times \mathcal{P}^*$$
$$= \{(s, (q, \sigma), P) \in \mathcal{J}^2 : s \leq 0, \ (q, \sigma) \in \mathcal{D}, \ and \ P \in \mathcal{P}^*\}, \quad (12.134)$$

where $\mathcal{P}^*$ *is defined by* (12.128). *Consequently, for every* $c \in F(\mathcal{J}^2)$ *and for every bounded domain* $\Omega \subset \mathbb{R}^n$, *one has the parabolic comparison principle*

$$u \leq w \text{ on } \partial^- \Omega \ \Rightarrow \ u \leq w \text{ on } \Omega \quad (12.135)$$

for $u \in \mathrm{USC}(\overline{\Omega})$ *and* $w \in \mathrm{LSC}(\overline{\Omega})$, *which are respectively* $\mathcal{F}_c$-*subharmonic and* $\mathcal{F}_c$-*superharmonic in* Ω *with respect to the subequation constraint set*

$$\mathcal{F}_c := \{(r, p, A) \in \mathcal{J}^2 : F(r, p, A) \geq c\}. \quad (12.136)$$

If one also requires that F *is topologically tame on* $\mathcal{F}$, *then the parabolic comparison principle* (12.135) *(equivalently) holds if* u *and* w *are respectively a viscosity subsolution and a viscosity supersolution to* $F(u, Du, D^2u) = c$ *on* Ω.

We note that the parabolic comparison principle (12.135) for $\mathcal{F}_c$ is equivalent to the statement

$$u + v \leq 0 \text{ on } \partial^- \Omega \ \Rightarrow \ u + v \leq 0 \text{ on } \Omega \tag{12.137}$$

for each pair $u, v \in \mathrm{USC}(\overline{\Omega})$ which are respectively $\mathcal{F}_c$ and $\widetilde{\mathcal{F}}_c$-subharmonic on Ω.

Proof of Theorem 12.37. One has that F is proper elliptic on all of $\mathcal{J}^2$ by (12.131) since, for each $(r, (p, \tau), A) \in \mathcal{J}^2$ and each $s \leq 0$ and $P \in \mathcal{P}$, one has

$$\begin{aligned}
F(r + s, (p, \tau), A + P) &= G(r + s, p, A' + P') - \tau \\
&\geq G(r, p, A') - \tau \\
&= F(r, (p, \tau), A).
\end{aligned}$$

For each $c \in F(\mathcal{J}^2)$, the set $\mathcal{F}_c$ defined by (12.136) is a subequation constraint set. It is clear that $\mathcal{F}_c$ is nonempty for $c \in F(\mathcal{J}^2)$ and closed by the continuity of F. The topological property (T) will follow from Proposition 4.7 once one establishes the $\mathcal{M}^*$-monotonicity of each $\mathcal{F}_c$ with respect to the monotonicity cone subequation $\mathcal{M}^*$. Moreover, since the cone $\mathcal{M}^*$ contains the minimal monotonicity set $\mathcal{M}_0 = \mathcal{N} \times \{0\} \times \mathcal{P}$, the negativity and positivity properties (N) and (P) for $\mathcal{F}_c$ follow from the $\mathcal{M}^*$-monotonicity, which does hold. Indeed, for $(r, (p, \tau), A) \in \mathcal{F}_c$ and $(s, (q, \sigma), P) \in \mathcal{M}^*$, then by (12.131) and (12.132) one has

$$\begin{aligned}
F(r + s, (p, \tau) + (q, \sigma), A + P) &= G(r + s, p + q, A' + P') - (\tau + \sigma) \\
&\geq G(r, p + q, A') - (\tau + \sigma) \\
&\geq G(r, p, A') - \tau - \sigma - \gamma|q| \\
&\geq G(r, p, A') - \tau \geq c,
\end{aligned}$$

since $(r, (p, \tau), A) \in \mathcal{F}_c$ and $(q, \sigma) \in \mathcal{D}$.

In preparation for the proof of the parabolic comparison principle in the form (12.137), notice that since

$$\mathcal{P} \subset \mathcal{P}^*, \quad \text{where } \mathcal{P}^* \text{ is defined by (12.128),}$$

$\mathcal{M}^* = \mathcal{N} \times \mathcal{D} \times \mathcal{P}^*$ contains the cone $\mathcal{M} = \mathcal{N} \times \mathcal{D} \times \mathcal{P}$ used in the more general situation of Theorem 12.36. Hence, $\widetilde{\mathcal{M}^*} \subset \widetilde{\mathcal{M}}$ and, since $\widetilde{\mathcal{M}}$-subharmonic functions satisfy the (ZMP), for each $z \in \widetilde{\mathcal{M}^*}(\Omega)$ (the set of $\mathrm{USC}(\overline{\Omega})$-functions which are $\widetilde{\mathcal{M}^*}$-subharmonic on Ω), one also has the (ZMP)

$$z \leq 0 \text{ on } \partial\Omega \ \Rightarrow \ z \leq 0 \text{ on } \Omega. \tag{12.138}$$

This will be used for the function $z := u + v + \varepsilon\varphi$, where $\varepsilon > 0$ will be chosen small and $\varphi \in \mathrm{USC}(\overline{\Omega})$ is defined by

$$\varphi(x,t) = -\frac{1}{T-t} \text{ for } t < T \quad \text{and} \quad \varphi(x,t) = -\infty \text{ for } t = T.$$

In Ω, where φ is smooth, one has

$$\varepsilon\varphi = -\frac{\varepsilon}{T-t} < 0, \quad D_t(\varepsilon\varphi) = -\frac{\varepsilon}{(T-t)^2} < 0, \quad D_x(\varepsilon\varphi) = 0, \quad \text{and} \quad D_x^2(\varepsilon\varphi) = 0,$$

so $J^2_{(x,t)}(\varepsilon\varphi) \in \mathcal{M}^*$ for each $(x,t) \in \Omega$ and hence $\varepsilon\varphi$ is $\mathcal{M}^*$-subharmonic in Ω.

Since the subequation $\mathcal{F}_c$ is $\mathcal{M}^*$-monotone, one has $u+v \in \widetilde{\mathcal{M}^*}(\Omega)$ by the subharmonic addition theorem (Theorem 7.4). Similarly, since $\mathcal{M}^* + \mathcal{M}^* \subset \mathcal{M}^*$ one also has

$$\mathcal{M}^* + \widetilde{\mathcal{M}^*} \subset \mathcal{M}^* \quad \text{and} \quad \mathcal{M}^*(\Omega) + \widetilde{\mathcal{M}^*}(\Omega) \subset \widetilde{\mathcal{M}^*}(\Omega).$$

Then, since $u+v \in \widetilde{\mathcal{M}^*}(\Omega)$ and $\varepsilon\varphi \in \mathcal{M}^*(\Omega)$, one has

$$z = (u+v) + \varepsilon\varphi \in \widetilde{\mathcal{M}^*}(\Omega), \tag{12.139}$$

and one can apply (12.138) to z.

Now, if the comparison principle (12.135) fails, there must be an interior point $(x_0, t_0) \in \Omega$ such that $(u+v)(x_0, t_0) > 0$. Since $u+v \leq 0$ on the parabolic boundary $\partial^- \Omega$ and since $\varphi(x,t) < 0$ for each $t < T$ and $\varphi(x,T) \equiv -\infty$, one has

$$z \leq 0 \text{ on } \partial\Omega \quad \text{and} \quad z(x_0, t_0) = (u+v)(x_0, t_0) - \frac{\varepsilon}{T-t} > 0$$

if $\varepsilon > 0$ is chosen sufficiently small. This contradicts the validity of (12.138). $\quad\square$

Next we present a parabolic version of the constrained case of Theorem 12.32. The proof is a simple adaptation of the proof of Theorem 12.37 and hence will be left to the reader.

Theorem 12.38. *Consider the operator defined by*

$$F(r, (p, \tau), A) = \tau G(r, A'), \tag{12.140}$$

with $G \colon \mathcal{G} \subsetneq \mathbb{R} \times \mathcal{S}(n-1) \to \mathbb{R}$ continuous and $\mathcal{G}$ closed and nonempty. Suppose that the pair $(G, \mathcal{G})$ is $\mathcal{N} \times \{0\} \times \mathcal{P}'$-monotone, that is,

$$(r, A') \in \mathcal{G} \implies (r+s, A' + P') \in \mathcal{G} \text{ for every } s \leq 0, \; P' \in \mathcal{P}', \tag{12.141}$$
$$G(r+s, A' + P') \geq G(r, A') \quad \text{for every } (r, A') \in \mathcal{G}, \; s \leq 0, \; P' \in \mathcal{P}'. \tag{12.142}$$

Suppose also that

$$\inf_{\mathcal{G}} G = 0 \quad and \quad \partial \mathcal{G} = \{(r, A') \in \mathcal{G} : G(r, A') = 0\}. \tag{12.143}$$

Then $(F, \mathcal{F})$ is a (constrained-case) compatible proper elliptic pair for the subequation

$$\mathcal{F} := \{(r, (p, \tau), A) \in \mathcal{J}^2 : (p, \tau) \in \mathcal{D}, \ (r, A') \in \mathcal{G}, \ F(r, (p, \tau), A) \geq 0\}, \tag{12.144}$$

where $\mathcal{D} \subsetneq \mathbb{R}^n$ is the directional cone (half-space)

$$\mathcal{D} := \{(p, \tau) \in \mathbb{R}^{n^-} \times \mathbb{R} : \tau \geq 0\}, \tag{12.145}$$

and the pair is $\mathcal{M}^*$-monotone for the monotonicity cone subequation

$$\mathcal{M}^* = \mathcal{N} \times \mathcal{D} \times \mathcal{P}^* := \{(s, q, P) \in \mathcal{J}^2 : s \leq 0, \ (q, \sigma) \in \mathcal{D}, \ P \in \mathcal{P}^*\}, \tag{12.146}$$

where $\mathcal{P}^*$ is defined by (12.128). Consequently, for every $c \in F(\mathcal{F})$ and for every bounded domain $\Omega \subset \mathbb{R}^n$, one has the parabolic comparison principle

$$u \leq w \ on \ \partial^- \Omega \ \Rightarrow \ u \leq w \ on \ \Omega \tag{12.147}$$

for $u \in \mathrm{USC}(\overline{\Omega})$ and $w \in \mathrm{LSC}(\overline{\Omega})$, which are respectively $\mathcal{F}_c$-subharmonic and $\mathcal{F}_c$-superharmonic in Ω, where $\mathcal{F}_c := \{(r, p, A) \in \mathcal{F} : F(r, p, A) \geq c\}$.

If one also requires that F is topologically tame on $\mathcal{F}$, then the comparison principle (12.147) equivalently holds if u and w are respectively an $\mathcal{F}$-admissible subsolution and an $\mathcal{F}$-admissible supersolution to $F(u, Du, D^2u) = c$ on Ω.

Appendix

Appendix A

Existence Holds and Uniqueness Implies Comparison

First, for the reader's convenience, we state a general result ([49, Theorem 12.7]) which in rough language says that, for any constant-coefficient subequation, "existence always holds" and "uniqueness bounds always hold" for the Dirichlet problem. Then we show how this can be used to prove that "uniqueness implies comparison," which was not stated in [49] but follows easily from this [49, Theorem 12.7].

Given a subequation $\mathcal{F}$ and a bounded domain Ω in Euclidean space $\mathbb{R}^n$, the Dirichlet problem for $\mathcal{F}$ on Ω can be formulated as follows.

Definition A.1. Given a boundary function $\varphi \in C(\partial\Omega)$, a *solution to the Dirichlet problem* (DP) *for $\mathcal{F}$ on Ω* is a function $h \in C(\overline{\Omega})$ satisfying

(1) $h|_{\Omega}$ is $\mathcal{F}$-harmonic on Ω;
(2) $h|_{\partial\Omega} = \varphi$.

Recalling that $\mathcal{F}(\overline{\Omega}) := \{u \in \mathrm{USC}(\overline{\Omega}) : u$ is $\mathcal{F}$-subharmonic on $\Omega\}$, the *associated Perron family* is defined by

$$\mathfrak{F}_{\mathcal{F},\varphi} := \left\{ u \in \mathcal{F}(\overline{\Omega}) : u|_{\partial\Omega} \leq \varphi \right\} \tag{A.1}$$

and the associated *Perron function* is defined on $\overline{\Omega}$ by

$$U_{\mathcal{F},\varphi}(x) := \sup\left\{ u(x) : u \in \mathfrak{F}_{\mathcal{F},\varphi} \right\}, \quad x \in \overline{\Omega}. \tag{A.2}$$

There are two natural invariants associated with a subequation $\mathcal{F}$. The first is the maximal monotonicity cone $\mathcal{M}_{\mathcal{F}}$ which was discussed in Section 4.1. The second is the *asymptotic interior* $\overrightarrow{\mathcal{F}}$, which is crucial for defining $\mathcal{F}$-*boundary convexity of Ω*. We refer the reader to [49, Section 11] for a discussion of this. Now we state the strong form of "existence always holds," where "always" refers to the fact that there is no restriction on the subequation $\mathcal{F} \subset \mathcal{J}^2$ if the boundary is suitably convex.

Theorem A.2. *Suppose that $\mathcal{F}$ is a constant-coefficient subequation and that $\Omega \subset\subset \mathbb{R}^n$ has a C^2-boundary which is both strictly $\overrightarrow{\mathcal{F}}$-convex and strictly*

$\vec{\mathcal{F}}$-convex. Suppose that a boundary function $\varphi \in C(\partial\Omega)$ is given. Then the following two statements hold:

(a) If a solution h to the (DP) of Definition A.1 exists, then h lies between minus the dual Perron function $-U_{\widetilde{\mathcal{F}},-\varphi}$ and the Perron function $U_{\mathcal{F},\varphi}$, that is,

$$-U_{\widetilde{\mathcal{F}},-\varphi} \le h \le U_{\mathcal{F},\varphi} \quad \text{on } \overline{\Omega}. \tag{A.3}$$

(b) The bounding functions $-U_{\widetilde{\mathcal{F}},-\varphi}$ and $U_{\mathcal{F},\varphi}$ always solve the (DP) of Definition A.1.

Corollary A.3. The negative $-U_{\widetilde{\mathcal{F}},-\varphi}$ of the Perron function $U_{\widetilde{\mathcal{F}},-\varphi}$ (which solves the (DP) for $\widetilde{\mathcal{F}}$ and boundary data $-\varphi$) also solves the (DP) for $\mathcal{F}$ and boundary data $\varphi \in C(\partial\Omega)$.

Proof. Apply Theorem A.2 to the subequation $\widetilde{\mathcal{F}}$ and the boundary function $-\varphi$. Then $U_{\widetilde{\mathcal{F}},-\varphi}$ satisfies

(1) $U_{\widetilde{\mathcal{F}},-\varphi}$ restricted to Ω is $\widetilde{\mathcal{F}}$-subharmonic and $-U_{\widetilde{\mathcal{F}},-\varphi}$ restricted to Ω is $\mathcal{F} = \widetilde{\widetilde{\mathcal{F}}}$-subharmonic Ω;
(2) $U_{\widetilde{\mathcal{F}},-\varphi}$ restricted to $\partial\Omega$ is $-\varphi$. $\square$

Corollary A.4. If $h \in C(\overline{\Omega})$ is any other solution of (DP) for $\mathcal{F}$ on Ω, then h lies between $-U_{\widetilde{\mathcal{F}},-\varphi}$ and $U_{\mathcal{F},\varphi}$, that is, with equality on $\partial\Omega$.

Proof. Note that h belongs to the Perron family $\mathfrak{F}_{\mathcal{F},\varphi}$ and hence $h \le U_{\mathcal{F},\varphi}$, while $-h$ belongs to the Perron family $\mathfrak{F}_{\widetilde{\mathcal{F}},-\varphi}$ and hence $-h \le U_{\widetilde{\mathcal{F}},-\varphi}$. $\square$

Now "uniqueness implies comparison" is straightforward.

Theorem A.5. Assume that the hypotheses of Theorem A.2 hold. Then if there is at most one solution to (DP) for $\mathcal{F}$ on Ω for each fixed boundary function $\varphi \in C(\overline{\Omega})$, it follows that comparison holds for $\mathcal{F}$ on $\overline{\Omega}$.

Proof. By Theorem A.2, Corollary A.3, and the uniqueness assumption one has

$$-U_{\widetilde{\mathcal{F}},-\varphi} = U_{\mathcal{F},\varphi} \quad \text{on } \overline{\Omega}. \tag{A.4}$$

Suppose that $u \in \mathcal{F}(\overline{\Omega})$ and $v \in \widetilde{\mathcal{F}}(\overline{\Omega})$ with $u + v \le 0$ on $\partial\Omega$.

Assume, for the moment, that at least one of the functions u or v belongs to $C(\overline{\Omega})$, say $u \in C(\overline{\Omega})$. Set $\varphi := u|_{\partial\Omega} \in C(\partial\Omega)$. Then u belongs to the Perron family $\mathfrak{F}_{\mathcal{F},\varphi}$ and hence

$$u \le U_{\mathcal{F},\varphi} \quad \text{on } \overline{\Omega}.$$

Now, the hypothesis $u + v \leq 0$ on $\partial\Omega$ is equivalent to $v|_{\partial\Omega} \leq -\varphi$. Hence, v belongs to the Perron family $\mathfrak{F}_{\widetilde{F},-\varphi}$ and thus

$$v \leq U_{\widetilde{F},-\varphi} \quad \text{on } \overline{\Omega}.$$

Therefore,

$$u + v \leq U_{\mathcal{F},\varphi} + U_{\widetilde{F},-\varphi} \quad \text{on } \overline{\Omega},$$

where this last sum is zero by (A.4), which used the assumption that uniqueness holds.

Finally, the provisional assumption that one of the functions u or v is continuous on $\overline{\Omega}$ can be removed. In fact, for fixed $u \in \mathrm{USC}(\overline{\Omega})$, given $\varepsilon > 0$ there exists $u_\varepsilon \in C(\overline{\Omega})$ such that

$$u \leq u_\varepsilon \leq u + \varepsilon \quad \text{on } \overline{\Omega}. \tag{A.5}$$

Take $\varphi_\varepsilon := u_\varepsilon|_{\partial\Omega} \in C(\partial\Omega)$ and consider the Perron function $U_{\mathcal{F},\varphi_\varepsilon}$. By the negativity property for $\mathcal{F}$ and (A.5) one has that

$$U_{\mathcal{F},\varphi_\varepsilon} - \varepsilon \in \mathcal{F}(\overline{\Omega}) \cap C(\overline{\Omega}) \quad \text{and} \quad U_{\mathcal{F},\varphi_\varepsilon} - \varepsilon = \varphi_\varepsilon - \varepsilon \leq u \text{ on } \partial\Omega. \tag{A.6}$$

Since $u + v \leq 0$ on $\partial\Omega$ by hypothesis, $U_{\mathcal{F},\varphi_\varepsilon} - \varepsilon + v \leq 0$ on $\partial\Omega$ (since we are assuming uniqueness also holds for the boundary function φ_ε), and by the previous step it follows that

$$U_{\mathcal{F},\varphi_\varepsilon} - \varepsilon + v \leq 0 \quad \text{on } \overline{\Omega},$$

but $u \leq U_{\mathcal{F},\varphi_\varepsilon}$ and $\varepsilon > 0$ is arbitrary. $\qquad\square$

Remark A.6. All of the results in this appendix have natural extensions from $\mathbb{R}^n$ to any Riemannian homogeneous space $X \equiv K/G$ with a subequation $\mathcal{F}$ which is invariant under the natural action of the Lie group K on the 2-jet bundle $J^2(X)$. See [49, Theorem 13.5].

The following result is useful in identifying the maximal solution $U_{\mathcal{F},\varphi}$ in (A.3), especially in examples. No boundary convexity or smoothness of $\partial\Omega$ is required.

Lemma A.7. *Suppose that h is a solution to the (DP) for $\mathcal{F}$ on Ω with boundary values $\varphi \in C(\partial\Omega)$. That is, $h \in C(\overline{\Omega})$, $h|_\Omega$ is $\mathcal{F}$-harmonic on Ω, and $h|_{\partial\Omega} = \varphi$. If h can be pointwise approximated by $\{h_j\}_{j\in\mathbb{N}} \subset C(\overline{\Omega}) \cap C^2(\Omega)$ with each $-h_j$ being strictly $\widetilde{\mathcal{F}}$-subharmonic on Ω, then h is the Perron function $U_{\mathcal{F},\varphi}$.*

Proof. Choose any u in the Perron family $\mathfrak{F}_{\mathcal{F},\varphi}$. We need to show that $u \leq h$ on $\overline{\Omega}$. To this end, set $c_j := \max\{0, \sup_{\partial\Omega}(\varphi - h_j)\}$ so that

$$0 \leq c_j \quad \text{and} \quad (\varphi - h_j)|_{\partial\Omega} \leq c_j.$$

Then, on $\partial\Omega$, $u - c_j - h_j \leq 0$. Since $c_j \geq 0$, by the negativity property (N) for $\mathcal{F}$, we have $u - c_j \in \mathcal{F}(\overline{\Omega})$. Since $-h_j$ is strictly $\widetilde{\mathcal{F}}$-subharmonic, we can apply definitional comparison (Lemma 3.14) to $(u - c_j)$ and $-h_j$ to conclude that

$$u - c_j - h_j \leq 0 \quad \text{on } \overline{\Omega}.$$

Since $c_j \to 0$ and $h_j \to h$ as $j \to +\infty$, we have $u - h \leq 0$ on $\overline{\Omega}$, as desired. $\qquad\square$

Of course, Lemma A.7 also identifies when a solution h to the (DP) for $\mathcal{F}$ with boundary data φ is the minimal solution $-U_{\widetilde{\mathcal{F}}, -\varphi}$ in (A.3), by identifying when $-h$ is the maximal solution to the (DP) for $\widetilde{\mathcal{F}}$ on Ω with boundary data $-\varphi$.

Corollary A.8. *Suppose that h is a solution to the* (DP) *for $\mathcal{F}$ on Ω with boundary values $\varphi \in C(\partial\Omega)$. If h can be pointwise approximated by $\{h_j\}_{j\in\mathbb{N}} \subset C(\overline{\Omega}) \cap C^2(\Omega)$, with each h_j being strictly $\mathcal{F}$-subharmonic on Ω, then*

$$h = -U_{\widetilde{\mathcal{F}}, -\varphi} \quad \text{on } \overline{\Omega}$$

is the minimal solution.

Moreover, if $-h$ can also be pointwise approximated by a sequence of functions in $C(\overline{\Omega}) \cap C^2(\Omega)$ which are strictly $\widetilde{\mathcal{F}}$-subharmonic on Ω, then

$$-U_{\widetilde{\mathcal{F}}, -\varphi} = h = U_{\mathcal{F}, \varphi} \quad \text{on } \overline{\Omega},$$

and hence comparison holds for $\mathcal{F}$ on Ω.

Appendix B

Failure of Comparison on Small Balls: Radial Proof

Making use of the considerations of Appendix A on maximal and minimal solutions, we will prove Theorem 9.8 concerning the failure of comparison on arbitrarily small balls for the (reduced) subequations

$$\mathcal{F} := \left\{ (p, A) \in \mathbb{R}^n \times \mathcal{S}(n) : \lambda_{\min}(B(p, A)) \geq 0 \right\} \tag{B.1}$$

and

$$\mathcal{G} := \left\{ (p, A) \in \mathbb{R}^n \times \mathcal{S}(n) : \lambda_{\max}(B(p, A)) \geq 0 \right\}, \tag{B.2}$$

where $\alpha \in (1, +\infty)$ is fixed and

$$B(p, A) := A + |p|^{\frac{\alpha-1}{\alpha}} (P_{p^\perp} + \alpha P_p) \quad \text{if } p \neq 0 \text{ and } B(0, A) := A. \tag{B.3}$$

Proof of Theorem 9.8. With $R \in (0, +\infty)$ and $B_R \subset \mathbb{R}^n$ the open R-ball about 0, it suffices to show that the C^2-functions defined by

$$z(x) := 0 \quad \text{and} \quad h(x) := -\frac{|x|^{1+\alpha}}{1+\alpha} + \frac{R^{1+\alpha}}{1+\alpha}, \quad x \in \mathbb{R}^n \tag{B.4}$$

are both $\mathcal{F}$- and $\mathcal{G}$-harmonic on all of $\mathbb{R}^n$. Obviously, they both take on the boundary values $\varphi = 0$ on ∂B_R.

For $z \equiv 0$, one has $(p, A) = (Dz, D^2z) \equiv (0, 0) \in \partial\mathcal{F} \cap \partial\mathcal{G}$ by (9.22).

Since $\alpha > 1$, h is $C^{2,\alpha-1}(\mathbb{R}^n)$. At the origin $(Dh(0), D^2h(0)) = (0, 0) \in \partial\mathcal{F} \cap \partial\mathcal{G}$. Hence, h is both $\mathcal{F}$- and $\mathcal{G}$-harmonic at the origin.

To calculate derivatives of h away from the origin, we use the fact that h is radial. It is enlightening to calculate the associated *radial subequation* $\mathcal{R}$ for $\mathcal{F}$, with dual $\widetilde{\mathcal{R}}$ the radial subequation for $\widetilde{\mathcal{F}}$ (leaving the analogous calculations for $\mathcal{G}$ to the reader) which apply even if h is not C^2.

Lemma B.1 (Radial subharmonics). *Suppose that $u \in C^2(\mathbb{R}^n)$ is radial with profile ψ, that is, $u(x) = \psi(|x|)$. Then u is $\mathcal{F}$-subharmonic on $\mathbb{R}^n \setminus \{0\}$ if and only if ψ is $\mathcal{R}$-subharmonic on $(0, +\infty)$, where $\mathcal{R}$ is defined by the conditions*

$$\frac{\psi'(t)}{t} + |\psi'(t)|^{\frac{\alpha-1}{\alpha}} \geq 0 \quad and \quad \psi''(t) + \alpha|\psi'(t)|^{\frac{\alpha-1}{\alpha}} \geq 0, \quad with\ t > 0. \qquad (B.5)$$

Similarly, u is $\widetilde{\mathcal{F}}$-subharmonic on $\mathbb{R}^n \setminus \{0\}$ if and only if ψ is $\widetilde{\mathcal{R}}$-subharmonic on $(0, +\infty)$, where $\widetilde{\mathcal{R}}$ is determined by the conditions

$$\frac{\psi'(t)}{t} - |\psi'(t)|^{\frac{\alpha-1}{\alpha}} \geq 0 \quad or \quad \psi''(t) - \alpha|\psi'(t)|^{\frac{\alpha-1}{\alpha}} \geq 0, \quad with\ t > 0. \qquad (B.6)$$

Proof. Using the radial calculation (3.24) of Remark 3.17, the reduced 2-jet of u is

$$(Du(x), D^2u(x)) = \left(\psi'(|x|)\frac{x}{|x|}, \frac{\psi'(|x|)}{|x|}P_{x^\perp} + \psi''(|x|)P_x\right),$$

where, as always, P_x and $P_{x^\perp}$ are the orthogonal projections onto the subspaces $[x]$ and $[x^\perp]$ for $x \neq 0$. Denoting $(p, A) := (Du(x), D^2u(x))$, one has

$$P_p = P_x, \quad P_{p^\perp} = P_{x^\perp}, \quad and \quad |p| = |\psi'(|x|)|,$$

as well as

$$A + |p|^{\frac{\alpha-1}{\alpha}}(P_{p^\perp} + \alpha P_p) = \left(\frac{\psi'(|x|)}{|x|} + |\psi'(|x|)|^{\frac{\alpha-1}{\alpha}}\right)P_{x^\perp}$$
$$+ (\psi''(|x|) + \alpha|\psi'(|x|)|^{\frac{\alpha-1}{\alpha}})P_x,$$

where this gives an element of $\mathcal{P}$ $(\lambda_{\min} \geq 0)$ if and only if ψ satisfies (B.5). Similarly,

$$A - |p|^{\frac{\alpha-1}{\alpha}}(P_{p^\perp} + \alpha P_p) = \left(\frac{\psi'(|x|)}{|x|} - |\psi'(|x|)|^{\frac{\alpha-1}{\alpha}}\right)P_{x^\perp}$$
$$+ (\psi''(|x|) - \alpha|\psi'(|x|)|^{\frac{\alpha-1}{\alpha}})P_x$$

will be an element of $\widetilde{\mathcal{P}}$ $(\lambda_{\max} \geq 0)$ if and only if ψ satisfies (B.6). $\qquad\square$

Remark B.2. Lemma B.1 also holds for $u \in \mathrm{USC}(\mathbb{R}^n)$, where we note that if $u(x) := \psi(|x|)$, then $u \in \mathrm{USC}(\mathbb{R}^n)$ if and only if $\psi \in \mathrm{USC}([0, +\infty))$. For the proof of Lemma B.1 for radial USC-functions, see [59, Section 2].

We now return to the proof of Theorem 9.8. Applying Lemma B.1 to the profile $\psi(t) = -\frac{t^{1+\alpha}}{1+\alpha}$, one computes to find

$$\frac{\psi'(t)}{t} + |\psi'(t)|^{\frac{\alpha-1}{\alpha}} = -\frac{t^\alpha}{t} + |-t^\alpha|^{\frac{\alpha-1}{\alpha}} = 0,$$
$$\psi''(t) + \alpha|\psi'(t)|^{\frac{\alpha-1}{\alpha}} = -\alpha t^{\alpha-1} + \alpha|-t^\alpha|^{\frac{\alpha-1}{\alpha}} = 0.$$

Note, in fact, that $A + |p|^{\frac{\alpha-1}{\alpha}}(P_{p^\perp} + P_p) = 0$ has all eigenvalues zero. Thus u is $\mathcal{F}$-harmonic on $\mathbb{R}^n \setminus \{0\}$ and so is h (which is smooth and differs from u by a constant). This completes the proof of Theorem 9.8. $\qquad\square$

We can also characterize these two distinct $\mathcal{F}$-harmonics as the extremals in Theorem A.2.

Proposition B.3. *For z and h as in Theorem 9.8 (and recalled in (B.4)), the following hold:*

(a) *The zero function z equals $-U_{\widetilde{\mathcal{F}},-\varphi}$, where $U_{\widetilde{\mathcal{F}},-\varphi}$ is the Perron function for $\widetilde{\mathcal{F}}$ with boundary data $\varphi = 0$.*
(b) *The function h equals $U_{\mathcal{F},-\varphi}$, where $U_{\mathcal{F},-\varphi}$ is the Perron function for $\mathcal{F}$ on B_R with boundary data $\varphi = 0$.*

Moreover, the same statements hold with $\mathcal{F}$ replaced by $\mathcal{G}$.

Proof. For both claims (a) and (b), we will use definitional comparison by way of Lemma A.7 and Corollary A.8.

For part (a), by Corollary A.8 it suffices to show that for each (small) $\varepsilon > 0$ the function defined by

$$z_\varepsilon(x) := z(x) + \frac{\varepsilon}{2}|x|^2 = \frac{\varepsilon}{2}|x|^2$$

is strictly $\mathcal{F}$-subharmonic (since z_ε is regular and pointwise approximates z). Now, $\psi_\varepsilon(t) = \frac{\varepsilon}{2}t^2$ defines $z_\varepsilon(x) = \psi_\varepsilon(|x|)$ with $\psi'_\varepsilon(t) = \varepsilon t$ and $\psi''_\varepsilon(t) = \varepsilon$, so that

$$\frac{\psi'_\varepsilon(t)}{t} + |\psi'_\varepsilon(t)|^{\frac{\alpha-1}{\alpha}} = \varepsilon + (\varepsilon t)^{\frac{\alpha-1}{\alpha}} > 0$$

and $\psi''_\varepsilon(t) + |\psi'_\varepsilon(t)|^{\frac{\alpha-1}{\alpha}}$ gives the same (positive) quantity. Thus we have the needed strict inequalities in (B.5). Said differently,

$$A + |p|^{\frac{\alpha-1}{\alpha}}(P_{p^\perp} + \alpha P_p) = (\varepsilon + (\varepsilon t)^{\frac{\alpha-1}{\alpha}})I \in \text{Int } \mathcal{P}.$$

Notice that since $\mathcal{F} \subset \mathcal{G}$, z_ε will also be strictly $\mathcal{G}$-subharmonic and we have part (a) for $\mathcal{G}$.

For part (b), since $\widetilde{\mathcal{G}} \subset \widetilde{\mathcal{F}}$, by Lemma A.7 and the form of (B.6), it suffices to find a $\widetilde{\mathcal{G}}$-strict C^2-approximation to $-h(x) + \frac{R^{1+\alpha}}{1+\alpha} = \frac{|x|^{1+\alpha}}{1+\alpha}$. Set $\psi(t) := \frac{t^{1+\alpha}}{1+\alpha}$ and define $\psi_\varepsilon(t) := \frac{(1+\varepsilon)(t+\varepsilon)^{1+\alpha}}{1+\alpha}$. It remains to show that $\psi_\varepsilon(t)$ is a strict subharmonic for the radial subequation associated to $\widetilde{\mathcal{G}}$, that is,

$$\frac{\psi'_\varepsilon(t)}{t} - |\psi'_\varepsilon(t)|^{\frac{\alpha-1}{\alpha}} > 0 \quad \text{and} \quad \frac{1}{\alpha}\psi''_\varepsilon(t) - |\psi'_\varepsilon(t)|^{\frac{\alpha-1}{\alpha}} > 0. \tag{B.7}$$

Now, $\psi'_\varepsilon(t) = (1+\varepsilon)(t+\varepsilon)^\alpha$ and

$$\frac{1}{\alpha}\psi''_\varepsilon(t) = (1+\varepsilon)(t+\varepsilon)^{\alpha-1} < \frac{\psi'_\varepsilon(t)}{t} = (1+\varepsilon)(t+\varepsilon)^{\alpha-1}\left(1+\frac{\varepsilon}{t}\right).$$

Hence, it suffices to verify the second inequality in (B.7), but

$$\frac{1}{\alpha}\psi''_\varepsilon(t) - |\psi'_\varepsilon(t)|^{\frac{\alpha-1}{\alpha}} = (1+\varepsilon)(t+\varepsilon)^{\alpha-1}\left(1 - \frac{1}{(1+\varepsilon)^{1/\alpha}}\right) > 0. \qquad \Box$$

Appendix C

Equivalent Definitions of $\mathcal{F}$-Subharmonic Functions

Here we include the elementary facts in [49, Appendix A] , but presented in a different manner, more closely related to the notion of a function being $\mathcal{F}$-subharmonic at a point.

There are at least four different possibilities for defining the space of (upper) test jets for u at x_0. Given a 2-jet $J = (r, p, A)$, let

$$Q_J(x) := r + \langle p, x - x_0 \rangle + \tfrac{1}{2} \langle A(x - x_0), x - x_0 \rangle \tag{C.1}$$

denote the quadratic function with 2-jet J at x_0.

Lemma C.1. *Suppose that $u \in \mathrm{USC}(X)$ and $x_0 \in X$. The following sets of test jets for u at x_0 all have the same closure in $\mathcal{J}^2$:*

(J1) Strict quadratic test jets:

$$J_1(x_0, u) = \big\{ J \in \mathcal{J}^2 : \text{for some } \varepsilon > 0, \; u(x) - Q_J(x) \leq -\varepsilon |x - x_0|^2$$
$$\text{near } x_0, \text{ with equality at } x = x_0 \big\}.$$

(J2) Quadratic test jets:

$$J_2(x_0, u) = \big\{ J \in \mathcal{J}^2 : u(x) - Q_J(x) \leq 0 \text{ near } x_0,$$
$$\text{with equality at } x = x_0 \big\}.$$

(J3) C^2-test jets:

$$J_3(x_0, u) = \big\{ J^2_{x_0} \varphi : \varphi \in C^2 \text{ near } x_0, \; u(x) - \varphi(x) \leq 0 \text{ near } x_0,$$
$$\text{with equality at } x = x_0 \big\}.$$

(J4) Little-*o* quadratic test jets:

$$J_4(x_0, u) = \left\{ J \in \mathcal{J}^2 : u(x) - Q_J(x) \leq o(|x - x_0|^2) \ near \ x_0, \right.$$
$$\left. with \ equality \ at \ x = x_0 \right\}.$$

Proof. First, note that

$$J_1(x_0, u) \subset J_2(x_0, u) \subset J_3(x_0, u) \subset J_4(x_0, u), \qquad (C.2)$$

and hence it suffices to show that

$$J_4(x_0, u) \subset \bar{J}_1(x_0, u). \qquad (C.3)$$

Suppose that $J \in J_4(x_0, u)$, that is,

$$u(x_0) - Q_J(x_0) = 0 \quad and \quad u(x) - Q_J(x) \leq o(|x - x_0|^2) \ as \ x \to x_0.$$

Hence, for each $\varepsilon > 0$ there exists a neighborhood $B_\delta(x_0)$ with $\delta = \delta(\varepsilon)$ such that

$$u(x) - Q_J(x) \leq \varepsilon |x - x_0|^2 \ in \ B_\delta(x_0) \quad and \quad u(x_0) - Q_J(x_0) = 0. \qquad (C.4)$$

Denoting $J_\alpha := J + (0, 0, \alpha I)$, since $-Q_{J_{4\varepsilon}}(x) = -Q_J(x) - 2\varepsilon |x - x_0|^2$, (C.4) can be written as

$$u(x) - Q_{J_{4\varepsilon}}(x) \leq -\varepsilon |x - x_0|^2 \ in \ B_\delta(x_0) \quad and \quad u(x_0) - Q_{J_{4\varepsilon}}(x_0) = 0. \qquad (C.5)$$

Hence, $J_{4\varepsilon} \in J_1(x_0, u)$ for each $\varepsilon > 0$ and taking the limit as $\varepsilon \to 0^+$ gives $J \in \bar{J}_1(x_0, u)$ and hence (C.3). $\qquad \square$

Corollary C.2. *Let $\mathcal{F} \subset \mathcal{J}^2$ be an arbitrary closed subset. Given $u \in \mathrm{USC}(X)$ and $x_0 \in X$, the conditions*

$$J_1(x_0, u) \subset \mathcal{F}, \quad J_2(x_0, u) \subset \mathcal{F}, \quad J_3(x_0, u) \subset \mathcal{F}, \quad and \quad J_4(x_0, u) \subset \mathcal{F} \qquad (C.6)$$

are all equivalent.

Proof. If $J_k(x_0, u) \subset \mathcal{F}$ for some $k \in \{1, 2, 3, 4\}$, then $\bar{J}_k \subset \mathcal{F}$ since $\mathcal{F}$ is closed, but $\bar{J}_j(x_0, u) = \bar{J}_k(x_0, u)$ for each $j \neq k$ by Lemma C.1. $\qquad \square$

Definition C.3. *Given $\mathcal{F}$ a subequation constraint set, a function $u \in \mathrm{USC}(X)$ is $\mathcal{F}$-subharmonic at $x_0 \in X$ if*

$$J \in \mathcal{F} \ for \ all \ test \ jets \ J \ for \ u \ at \ x_0, \qquad (C.7)$$

where one may adopt any of the four definitions of (upper) test jets for u at x_0 contained in Lemma C.1.

As noted in Chapter 2, Corollary C.2 shows that the equivalent ways of defining $\mathcal{F}$-subharmonicity in x_0 do not depend on the subequation constraint properties (P) and (N), but only on the fact that $\mathcal{F}$ is closed, which follows from property (T), which also ensures that $\mathcal{F}$ is nonempty, and hence the conditions (C.6) are nontrivial.

Elementary Properties of $\mathcal{F}$-Subharmonic Functions

We consider $\mathcal{F}$-subharmonic functions on an open set $X \subset \mathbb{R}^n$ with $\mathcal{F} \subset \mathcal{J}^2$ a subequation constraint set (see Definitions 2.1 and 2.4). While these are known properties (see [46, Theorem 2.6 and Appendix B]), for the convenience of the reader we reproduce the proofs here, making use of Lemma C.1 which has been tailored to the pointwise notion of $\mathcal{F}$-subharmonicity.

Proposition D.1 (Elementary properties of $\mathcal{F}(X)$). *For $\mathcal{F}$ and X as above, the following hold:*

(A) (Local property). $u \in \mathrm{USC}(X)$ *is locally $\mathcal{F}$-subharmonic* $\Leftrightarrow$ $u \in \mathcal{F}(X)$.
(B) (Maximum property). *If $u, v \in \mathcal{F}(X)$ then $\max\{u, v\} \in \mathcal{F}(X)$.*
(C) (Coherence property). *If $u \in \mathrm{USC}(X)$ is twice differentiable in $x_0 \in X$, then u is $\mathcal{F}$-subharmonic in x_0 if and only if $J^2_{x_0} u \in \mathcal{F}$.*
(D) (Translation property). $u \in \mathcal{F}(X)$ $\Leftrightarrow$ $u_y \in \mathcal{F}(X + y)$, *where $u_y(x) := u(x - y)$.*
(E) (Decreasing sequence property). *If $\{u_k\}_{k \in \mathbb{N}} \subset \mathcal{F}(X)$ is a decreasing sequence of functions, then the limit $u := \lim_{k \to \infty} u_k \in \mathcal{F}(X)$.*
(F) (Uniform limits property). *If $\{u_k\}_{k \in \mathbb{N}} \subset \mathcal{F}(X)$ is a sequence of functions which converges uniformly to u on compact subsets of X, then $u \in \mathcal{F}(X)$.*
(G) (Families locally bounded above). *Suppose $\mathfrak{F} \subset \mathcal{F}(X)$ is a family of functions which are locally uniformly bounded above. Then the upper-semicontinuous regularization u^* of the upper envelope*

$$u(x) = \sup_{v \in \mathfrak{F}} v(x)$$

belongs to $\mathcal{F}(X)$.

Proof. (A) This is built into Definition 2.4 where locally $\mathcal{F}$-subharmonic just means that for each $x_0 \in X$, u is $\mathcal{F}$-subharmonic on some neighborhood of x_0.

(B) The condition that $\max\{u, v\} - \varphi \leq 0$ near x_0 with equality at x_0 implies that for one of the functions u, v, say u, we have $u(x_0) = \varphi(x_0)$. In this case, $u - \varphi \leq 0$ near x_0 with equality at x_0. Hence, $J^2_{x_0} \varphi \in \mathcal{F}$.

(C) This is the content of Remark 2.9, which makes use of the little-o quadratic jet formulation (J4) of Lemma C.1 for one direction and property (P) for the other.

(D) This is obvious since the fibers $\mathcal{F}_{x_0}$ do not depend on $x_0 \in \mathbb{R}^n$.

The remaining properties are all proved by contradiction with the fact that $\mathcal{F}$ is closed, using the bad test jet lemma (Lemma 2.8, which comes from negating the strict quadratic jet formulation (J1) of Lemma C.1). More precisely, by Lemma 2.8, if $u \in \mathrm{USC}(X)$ is not $\mathcal{F}$-subharmonic, then there exist $x_0 \in X$, $\varepsilon > 0$, $\rho > 0$, and a bad test jet $J = (r, p, A) \notin \mathcal{F}$ such that

$$u(x) - Q_J(x) \text{ satisfies } \begin{cases} \leq -\varepsilon |x - x_0|^2 & \text{on } B_\rho(x_0) \subset X, \\ = 0 & \text{at } x_0, \end{cases} \tag{D.1}$$

where $Q_J(x) = r + \langle p, x - x_0 \rangle + \frac{1}{2} \langle A(x - x_0), x - x_0 \rangle$ has $J^2_{x_0} Q_J = J = (r, p, A) \notin \mathcal{F}$ with $r = u(x_0)$. Given a bad test jet J for u at $x_0 \in X$, the idea is to exhibit a sequence of upper test jets $\{J_k = (r_k, p_k, A_k)\}_{k \in \mathbb{N}}$ (for a suitable sequence of $\mathcal{F}$-subharmonic functions) for which

$$\{J_k\}_{k \in \mathbb{N}} \subset \mathcal{F} \quad \text{and} \quad \lim_{k \to +\infty} J_k = J \notin \mathcal{F}, \tag{D.2}$$

a contradiction with $\mathcal{F}$ being closed.

(E) We begin by recalling that if $\{u_k\}_{k \in \mathbb{N}} \subset \mathrm{USC}(X)$ is a decreasing sequence, then the limit $u = \inf_{k \in \mathbb{N}} u_k$ is automatically $\mathrm{USC}(X)$. Hence, if $u \in \mathrm{USC}(X)$ is not $\mathcal{F}$-subharmonic, then the bad test jet lemma applies to u and there exist $x_0 \in X$, $\varepsilon > 0$, $\rho > 0$, and $J = (r, p, A) \notin \mathcal{F}$ such that (D.1) holds. By reducing $\rho > 0$ if necessary, we can assume that the compact neighborhood $\overline{B}_\rho(x_0)$ is contained in the open set X and then (D.1) shows that the upper-semicontinuous function $u - Q_J$ has a unique strict maximum on $\overline{B}_\rho(x_0)$ at $x = x_0$ with $u(x_0) - Q_J(x_0) = 0$.

We will construct the needed sequence of test jets (satisfying (D.2)) from the decreasing sequence of functions $\{v_k\}_{k \in \mathbb{N}} \subset \mathrm{USC}(X)$ defined by

$$v_k(x) := u_k(x) - Q_J(x), \tag{D.3}$$

which have nonnegative maxima

$$m_k := \sup_{\overline{B}_\rho(x_0)} v_k = \sup_{\overline{B}_\rho(x_0)} (u_k - Q_J) \geq \sup_{\overline{B}_\rho(x_0)} (u - Q_J) = u(x_0) - Q_J(x_0) = 0. \tag{D.4}$$

We recall the elementary fact that, for any decreasing sequence $\{v_k\}_{k \in \mathbb{N}} \subset \mathrm{USC}(X)$ with limit v and for any compact subset K of X, one has

$$\lim_{k \to +\infty} \left\{ \sup_K v_k \right\} = \sup_K v \quad \text{in } \mathbb{R}. \tag{D.5}$$

To see this, let $\alpha > 0$ be arbitrary and consider the sequence of sets $\{K_k\}_{k \in \mathbb{N}}$ defined by

$$K_k := \{x \in K : v_k(x) \geq \sup_K v + \alpha\}.$$

Each K_k is compact by the upper semicontinuity of v_k and $\{K_k\}_{k \in \mathbb{N}}$ is a decreasing sequence of sets ($K_{k+1} \subset K_k$ for each $k \in \mathbb{N}$) since $\{v_k\}_{k \in \mathbb{N}}$ is a decreasing sequence. By the pointwise convergence of v_k to v, one must have that $\bigcap_{k \in \mathbb{N}} K_k = \emptyset$ and hence each K_k must be empty for large k. More precisely, there exists $k_0 = k_0(\alpha) \in \mathbb{N}$ such that, for each $k \geq k_0$, one has $v_k(x) < \sup_K v + \alpha$ for every $x \in K$, and hence

$$\sup_K v \leq \sup_K v_k \leq \sup_K v + \alpha \quad \text{for every } k \geq k_0(\alpha),$$

which yields (D.5) since $\alpha > 0$ is arbitrary.

Returning to the construction, let $\delta \in (0, \rho)$ and consider the compact annulus $K_\delta := \overline{B}_\rho(x_0) \setminus B_\delta(x_0)$. By (D.1) one has $u(x) - Q_J(x) < 0$ on K_δ for each $\delta \in (0, \rho)$, and then applying (D.5) to the sequence $\{v_k\}_{k \in \mathbb{N}}$ in (D.3) on K_δ shows that there exists $k_1 = k_1(\delta) \in \mathbb{N}$ such that

$$\sup_{K_\delta}(u_k(x) - Q_J(x)) < 0 \quad \text{for every } k \geq k_1. \tag{D.6}$$

Consequently, for $k \geq k_1$, the nonnegative maximum m_k of (D.4) can only occur at points $x_k \in B_\rho(x_0)$. However, since δ can be made arbitrarily small there is a sequence $\{x_k\}_{k \geq k_1}$ such that

$$0 \leq m_k := \sup_{\overline{B}_\rho(x_0)} (u_k - Q_J) = u_k(x_k) - Q_J(x_k) \quad \text{and} \quad \lim_{k \to +\infty} x_k = x_0. \tag{D.7}$$

Hence, for $k \geq k_1$ one has

$$u_k(x) - (Q_J(x) + m_k) \leq 0 \text{ on } B_\rho(x_0) \quad \text{and} \quad u_k(x_k) - (Q_J(x_k) + m_k) = 0,$$

so that $Q_j + m_k$ is an upper test function for $u \in \mathcal{F}(X)$ at x_k and hence

$$J_k = (r_k, p_k, A_k) = D^2_{x_k}(Q_J + m_k) = (r + m_k, p + A(x - x_k), A) \in \mathcal{F}, \tag{D.8}$$

where $J = (r, p, A) \notin \mathcal{F}$ is the bad test jet with $r = u(x_0)$. Taking the limit as $k \to +\infty$, one has $x_k \to x_0$ by construction (D.7) and also $m_k \to 0$ by (D.5) and (D.4); that is,

$$\lim_{k \to +\infty} \sup_{\overline{B}_\rho(x_0)} (u_k - Q_J) = \sup_{\overline{B}_\rho(x_0)} (u - Q_J) = 0.$$

Hence, one has $\{J_k\}_{k \in \mathbb{N}} \subset \mathcal{F}$ with $J_k \to J \notin \mathcal{F}$, which contradicts $\mathcal{F}$ being closed.

(F) For uniform limits, the proof is almost the same as that given above for decreasing limits. In particular, using the definition of uniform convergence it is easy to see that the limit u is upper semicontinuous on X (so that Lemma 2.8 applies to give a bad test jet $J \notin \mathcal{F}$ at some point $x_0 \in X$) and that for each compact subset K of X the limit property (D.5),

$$\lim_{k \to +\infty} \left\{ \sup_K v_k \right\} = \sup_K v \quad \text{in } \mathbb{R},$$

holds if $v_k := u_k + Q_J$ converges uniformly to $u - Q_J$ on K. The same construction of the sequence of jets $\{J_k\}_{k \in \mathbb{N}} \subset \mathcal{F}$ with $J_k \to J \notin \mathcal{F}$ carries over without change.

(G) We give an adaptation of the classical proof in [30] by contradiction. Suppose that $u^* \notin \mathcal{F}(X)$. In order to simplify notation, we will assume that u^* is not $\mathcal{F}$-subharmonic in $x_0 = 0$ (which we may assume by the translation property (D)). By the bad test jet lemma (Lemma 2.8), there exist $\varepsilon > 0$, $\rho > 0$, and $(p, A) \in \mathbb{R}^n \times \mathcal{S}(N)$ such that

$$u^*(x) - [\langle p, x \rangle + \tfrac{1}{2}\langle Ax, x \rangle] \leq u^*(0) - \varepsilon |x|^2 \quad \text{for } |x| \leq \rho, \tag{D.9}$$

but

$$(u^*(0), p, A) \notin \mathcal{F}.$$

Since

$$u^*(0) = \lim_{k \to \infty} \sup_{|y| \leq \frac{1}{k}} \sup_{v \in \mathfrak{F}} v(y),$$

it follows easily that there exist sequences $\{y_k\} \subset \mathbb{R}^n$, $\{u_k\} \subset \mathfrak{F}$, such that $y_k \to 0$ and

$$\lim_{k \to \infty} u_k(y_k) = u^*(0). \tag{D.10}$$

Now choose a maximum point x_k for the function $u_k(x) - [\langle p, x \rangle + \tfrac{1}{2}\langle Ax, x \rangle]$ on $|x| \leq \rho$. Then

$$u_k(y_k) - [\langle p, y_k \rangle + \tfrac{1}{2}\langle Ay_k, y_k \rangle] \leq u_k(x_k) - [\langle p, x_k \rangle + \tfrac{1}{2}\langle Ax_k, x_k \rangle].$$

Pick a subsequence if necessary so that $x_k \to x$. Taking $\liminf_{k \to +\infty}$ of both sides and using the standard fact that $\limsup_{k \to +\infty} u_k(x_k) \leq u^*(x)$ yields

$$u^*(0) \leq \liminf_{k \to +\infty} u_k(x_k) - [\langle p, x \rangle + \tfrac{1}{2}\langle Ax, x \rangle] \quad \text{(by (D.10))}$$

$$\leq u^*(x) - [\langle p, x \rangle + \tfrac{1}{2}\langle Ax, x \rangle]$$

$$\leq u^*(0) - \varepsilon |x|^2 \quad \text{(by (D.9))}.$$

Consequently, we have $x = 0$ and

$$u^*(0) \leq \liminf_{k \to +\infty} u_k(x_k).$$

Since $x_k \to 0$, the upper semicontinuity of u^* gives

$$\limsup_{k \to +\infty} u_k(x_k) \leq u^*(0)$$

and therefore

$$\lim_{k \to \infty} u_k(x_k) = u^*(0).$$

The inequality

$$u_k(x) - [\langle p, x \rangle + \tfrac{1}{2}\langle Ax, x \rangle] \leq u_k(x_k) - [\langle p, x_k \rangle + \tfrac{1}{2}\langle Ax_k, x_k \rangle]$$

for $|x| \leq \rho$ (and hence near x_k) implies fairly easily that

$$(u_k(x_k), p + Ax_k, A) \in \mathcal{F}$$

since $u_k \in \mathcal{F}(X)$. Taking the limit as $k \to +\infty$ yields

$$(u^*(0), p, A) \in \mathcal{F},$$

a contradiction.   □

Bibliography

[1] M. E. Amendola, G. Galise, and A. Vitolo, *Riesz capacity, maximum principle, and removable sets of fully nonlinear second-order elliptic operators*, Differential Integral Equations **26** (2013), 845–866.

[2] D. J. Araújo, L. Mari, and L. F. Pessoa, *Detecting the completeness of a Finsler manifold via potential theory for its infinity Laplacian*, J. Differential Equations **281** (2021), 550–587.

[3] M. Bardi and I. Capuzzo Dolcetta, *Optimal control and viscosity solutions of Hamilton-Jacobi-Bellman equations*, Birkhäuser, Boston, 1997.

[4] M. Bardi and F. Da Lio, *On the strong maximum principle for fully nonlinear degenerate elliptic equations*, Arch. Math. (Basel) **73** (1999), 276–285.

[5] M. Bardi and P. Mannucci, *On the Dirichlet problem for non-totally degenerate fully nonlinear elliptic equations*, Commun. Pure Appl. Anal. **5** (2006), 709–731.

[6] G. Barles and J. Busca, *Existence and comparison results for fully nonlinear degenerate elliptic equations without zeroth-order term*, Comm. Partial Differential Equations **26** (2001), 2323–2337.

[7] I. Birindelli, G. Galise, and H. Ishii, *A family of degenerate elliptic operators: Maximum principle and its consequences*, Ann. Inst. H. Poincarè Anal. Non Linèaire **35** (2018), 417–441.

[8] I. Birindelli, G. Galise, and H. Ishii, *Towards a reversed Faber-Krahn inequality for the truncated Laplacian*, Rev. Mat. Iberoam. **36** (2020), 723–740.

[9] I. Birindelli, G. Galise, and H. Ishii, *Positivity sets of supersolutions of degenerate elliptic equations and the strong maximum principle*, Trans. Amer. Math. Soc. **374** (2021), 539–564.

[10] I. Birindelli and K. R. Payne, *Principal eigenvalues for k-Hessian operators by maximum principle methods*, Math. Eng. **3** (2021), Paper No. 021, 37 pp.

[11] J. M. Bony, *Principe du maximum, inégalité de Harnack et unicité du problème de Cauchy pour les opérateurs elliptiques dégénérées*, Ann. Inst. Fourier (Grenoble) **19** (1969), 277–304.

[12] K. K. Brustad, *On the comparison principle for second order elliptic equations without first and zeroth order terms*, NoDEA Nonlinear Dierential Equations Appl. **30** (2023), no. 1, Paper No. 15, 36 pp.

[13] K. K. Brustad, *Counterexamples to the comparison principle in the special Lagrangian potential equation*, arXiv:2206.09373v1, 2022.

[14] L. Caffarelli and X. Cabré, *Fully nonlinear elliptic equations*, American Mathematical Society Colloquium Publications 43, American Mathematical Society, Providence, RI, 1995.

[15] L. Caffarelli, L. Nirenberg, and J. Spruck, *The Dirichlet problem for nonlinear second-order elliptic equations III. Functions of the eigenvalues of the Hessian*, Acta Math. **155** (1985), 261–301.

[16] I. Capuzzo Dolcetta, F. Leoni, and A. Vitolo, *Entire subsolutions of fully nonlinear degenerate elliptic equations*, Bull. Inst. Math. Acad. Sin. (N.S.) **9** (2014), 147–161.

[17] J. Chen, M. Warren, and Y. Yuan, *A priori estimate for convex solutions to special Lagrangian equations and its application*, Comm. Pure Appl. Math. **62** (2009), 583–595.

[18] X. X. Chen and J. R. Cheng, *On the constant scalar curvature Kähler metrics. I: A priori estimates*, J. Amer. Math. Soc. **34** (2021), 909–936.

[19] X. Chen, S. Donaldson, and S. Sun, *Kähler-Einstein metrics on Fano manifolds. I: Approximation of metrics with cone singularities*, J. Amer. Math. Soc. **28** (2015), 183–197.

[20] X. Chen, S. Donaldson, and S. Sun, *Kähler-Einstein metrics on Fano manifolds. II: Limits with cone angle less than 2π*, J. Amer. Math. Soc. **28** (2015), 199–234.

[21] X. Chen, S. Donaldson, and S. Sun, *Kähler-Einstein metrics on Fano manifolds. III: Limits as cone angle approaches 2π and completion of the main proof*, J. Amer. Math. Soc. **28** (2015), 235–278.

[22] S. Y. Cheng and S. T. Yau, *Complete affine hypersurfaces. Part I. The completeness of affine metrics*, Comm. Pure Appl. Math. **39** (1986), 839–866.

[23] M. Cirant and K. R. Payne, *On viscosity solutions to the Dirichlet problem for elliptic branches of nonhomogeneous fully nonlinear equation*, Publ. Mat. **61** (2017), 529–575.

[24] M. Cirant and K. R. Payne, *Comparison principles for viscosity solutions of elliptic branches of fully nonlinear equations independent of the gradient*, Math. Eng. **3** (2021), Paper No. 045, 45 pp.

[25] M. Cirant, K. R. Payne, and D. F. Radaelli, *Comparison principles for nonlinear potential theory and PDEs with fiberegularity and sufficient monotonicity*, arXiv:2303.16735v1, 2023.

[26] A. Clarke and G. Smith, *The Perron method and the non-linear Plateau problem*, Geom. Dedicata **163** (2013), 159–164.

[27] T. C. Collins and S. T. Yau, *Moment maps, nonlinear PDE and stability in mirror symmetry. I: Geodesics*, Ann. PDE **7** (2021), Paper No. 11, 73 pp.

[28] M. G. Crandall, *Semidifferentials, quadratic forms and fully nonlinear elliptic equations of second order*, Ann. Inst. H. Poincaré C. Anal. Non Linéaire **6** (1989), 419–435.

[29] M. G. Crandall and H. Ishii, *The maximum principle for semicontinuous functions*, Differential Integral Equations **3** (1990), 1001–1014.

[30] M. G. Crandall, H. Ishii, and P.-L. Lions, *User's guide to viscosity solutions of second order partial differential equations*, Bull. Amer. Math. Soc. **27** (1992), 1–67.

[31] M. G. Crandall and P.-L. Lions, *Viscosity solutions of Hamilton-Jacobi equations*, Trans. Amer. Math. Soc. **277** (1983), 1–42.

[32] G. De Philippis and A. Figalli, *The Monge-Ampère equation and its link to optimal transportation*, Bull. Amer. Math. Soc. (N.S.) **51** (2014), 527–580.

[33] J. P. Demailly and N. Pali, *Degenerate complex Monge-Ampère equations over compact Kähler manifolds*, Internat. J. Math. **21** (2010), 357–405.

[34] L. C. Evans, *A convergence theorem for solutions of nonlinear second-order elliptic equations*, Indiana Univ. Math. J. **27** (1978), 875–887.

[35] L. C. Evans, *On solving certain nonlinear partial differential equations by accretive operator methods*, Israel J. Math. **36** (1980), 225–247.

[36] L. C. Evans, *The perturbed test function method for viscosity solutions of nonlinear PDE*, Proc. Roy. Soc. Edinburgh Sect. A **111** (1989), 359–375.

[37] L. C. Evans, *Periodic homogenisation of certain fully nonlinear partial differential equations*, Proc. Roy. Soc. Edinburgh Sect. A **120** (1992), 245–265.

[38] P. Eyssidieux, V. Guedj, and A. Zeriahi, *Singular Kähler-Einstein metrics*, J. Amer. Math. Soc. **22** (2009), 607–639.

[39] G. Galise and A. Vitolo, *Removable singularities for degenerate elliptic Pucci operators*, Adv. Differential Equations **22** (2017), 77–100.

[40] L. Gårding, *An inequality for hyperbolic polynomials*, J. Math. Mech. **8** (1959), 957–965.

[41] D. Gilbarg and N. S. Trudinger, *Elliptic partial differential equations of second order*, 2nd edition, Grundlehren der Mathematischen Wissenschaften 224, Springer, Berlin, 1983.

[42] A. Goffi and F. Pediconi, *A note on the strong maximum principle for fully nonlinear equations on Riemannian manifolds*, J. Geom. Anal. **31** (2021), 8641–8665.

[43] B. Guo, D. P. Phong, and F. Tong, *On L^∞ estimates for complex Monge-Ampère equations*, arXiv:2106.02224, 2021.

[44] F. R. Harvey and H. B. Lawson, Jr., *Calibrated geometries*, Acta Math. **148** (1982), 47–157.

[45] F. R. Harvey and H. B. Lawson, Jr., *An introduction to potential theory in calibrated geometry*, Amer. J. Math. **131** (2009), 893–944.

[46] F. R. Harvey and H. B. Lawson, Jr., *Dirichlet duality and the nonlinear Dirichlet problem*, Comm. Pure Appl. Math. **62** (2009), 396–443.

[47] F. R. Harvey and H. B. Lawson, Jr., *Duality of positive currents and plurisubharmonic functions in calibrated geometry*, Amer. J. Math. **131** (2009), 1211–1239.

[48] F. R. Harvey and H. B. Lawson, Jr., *Hyperbolic polynomials and the Dirichlet problem*, arXiv:0912.5220v2, 2010.

[49] F. R. Harvey and H. B. Lawson, Jr., *Dirichlet duality and the nonlinear Dirichlet problem on Riemannian manifolds*, J. Differential Geom. **88** (2011), 395–482.

[50] F. R. Harvey and H. B. Lawson, Jr., *Existence, uniqueness and removable singularities for nonlinear partial differential equations in geometry*, 103–156, in Surveys in Differential Geometry 18, International Press, Somerville MA, 2013.

[51] F. R. Harvey and H. B. Lawson, Jr., *Gårding's theory of hyperbolic polynomials*, Comm. Pure Appl. Math. **66** (2013), 1102–1128.

[52] F. R. Harvey and H. B. Lawson, Jr., *p-convexity, p-plurisubharmonicity and the Levi problem*, Indiana Univ. Math. J. **62** (2013), 149–170.

[53] F. R. Harvey and H. B. Lawson, Jr., *The restriction theorem for fully nonlinear subequations*, Ann. Inst. Fourier (Grenoble) **64** (2014), 217–265.

[54] F. R. Harvey and H. B. Lawson, Jr., *Potential theory on almost complex manifolds*, Ann. Inst. Fourier (Grenoble) **65** (2015), 171–210.

[55] F. R. Harvey and H. B. Lawson, Jr., *Characterizing the strong maximum principle for constant coefficient subequations*, Rend. Mat. Appl. (7) **37** (2016), 63–104.

[56] F. R. Harvey and H. B. Lawson, Jr., *Notes on the differentiation of quasi-convex functions*, arXiv:1309.1772v3, 2016.

[57] F. R. Harvey and H. B. Lawson, Jr., *The AE theorem and addition theorems for quasi-convex functions*, arXiv:1309.1770v3, 2016.

[58] F. R. Harvey and H. B. Lawson, Jr., *Lagrangian potential theory and a Lagrangian equation of Monge-Ampère type*, 217–258, in Surveys in Differential Geometry 22, International Press, Somerville, 2017.

[59] F. R. Harvey and H. B. Lawson, Jr., *Tangents to subsolutions: Existence and uniqueness, Part I*, Ann. Fac. Sci. Toulouse Math. (6) **27** (2018), 777–848.

[60] F. R. Harvey and H. B. Lawson, Jr., *The inhomogeneous Dirichlet problem for natural operators on manifolds*, Ann. Inst. Fourier (Grenoble) **69** (2019), 3017–3064.

[61] F. R. Harvey and H. B. Lawson, Jr., *A generalization of PDE's from a Krylov point of view*, Adv. Math. **372** (2020), Article ID 107298, 39 pp.

[62] F. R. Harvey and H. B. Lawson, Jr., *Pseudoconvexity for the special Lagrangian potential equation*, Calc. Var. Partial Differential Equations **60** (2021), Paper No. 6, 37 pp.

[63] F. R. Harvey, H. B. Lawson, Jr., and S. Pliś, *Smooth approximation of plurisubharmonic functions on almost complex manifolds*, Math. Ann. **366** (2016), 929–940.

[64] F. R. Harvey, H. B. Lawson, Jr., and S. Pliś, *The Richberg technique for subsolutions*, Comm. Anal. Geom. **28** (2020), 1787–1806.

[65] F. R. Harvey and K. R. Payne, *Interplay between nonlinear potential theory and fully nonlinear PDEs*, Pure Appl. Math. Q., to appear.

[66] H. Ishii, *Perron's method for Hamilton-Jacobi equations*, Duke Math. J. **55** (1987), 369–384.

[67] H. Ishii, *On uniqueness and existence of viscosity solutions of fully nonlinear second-order elliptic PDEs*, Comm. Pure Appl. Math. **42** (1989), 15–45.

[68] H. Ishii and P.-L. Lions, *Viscosity solutions of fully nonlinear second-order elliptic partial differential equations*, J. Differential Equations **83** (1990), 26–78.

[69] R. Jensen, *The maximum principle for viscosity solutions of fully nonlinear second order partial differential equations*, Arch. Rational Mech. Anal. **101** (1988), 1–27.

[70] R. Jensen, *Uniqueness criteria for viscosity solutions of fully nonlinear elliptic partial differential equations*, Indiana Univ. Math. J. **38** (1989), 629–667.

[71] R. Jensen, P.-L. Lions, and P. E. Souganidis, *A uniqueness result for viscosity solutions of second order fully nonlinear partial differential equations*, Proc. Amer. Math. Soc. **102** (1988), 975–978.

[72] S. Koike, *A beginner's guide to the theory of viscosity solutions*, MSJ Memoirs 13, Mathematical Society of Japan, Tokyo, 2004.

[73] S. Kolodziej, *The complex Monge-Ampère equation*, Acta Math. **180** (1998), 69–117.

[74] S. N. Kružkov, *First order quasilinear equations with several independent variables* (Russian) Mat. Sb. (N.S.) **81** (**123**) (1970), 228–255.

[75] N. V. Krylov, *Sequences of convex functions, and estimates of the maximum of the solution of a parabolic equation*, Siberian Math. J. **17** (1976), 226–236.

[76] N. V. Krylov, *Nonlinear elliptic and parabolic equations of second order*, Mathematics and Its Applications (Soviet Series) 7, D. Reidel, Dordrecht, 1987.

[77] N. V. Krylov, *On the general notion of fully nonlinear second-order elliptic equations*, Trans. Amer. Math. Soc. **347** (1995), 857–895.

[78] N. V. Krylov, *Sobolev and viscosity solutions for fully nonlinear elliptic and parabolic equations*, Mathematical Surveys and Monographs 233, American Mathematical Society, Providence, RI, 2018.

[79] P.-L. Lions, *Fully nonlinear elliptic equations and applications* 126–149, in *Nonlinear analysis, function spaces and applications, Vol. 2 (Písek, 1982)*, Teubner-Texte zur Mathematik 49, Teubner, Leipzig, 1982.

[80] P.-L. Lions, *Optimal control of diffusion processes and Hamilton-Jacobi-Bellman equations. II. Viscosity solutions and uniqueness*, Comm. Partial Differential Equations **8** (1983), 1229–1276.

[81] L. Mari and L. F. Pessoa, *Duality between Ahlfors-Liouville and Khas'min-skii properties for non-linear equations*, Comm. Anal. Geom. **28** (2020), 395–497.

[82] N. Nadirashvili and S. Vlăduţ, *Singular solution to special Lagrangian equations*, Ann. Inst. H. Poincaré C Anal. Non Linéaire **27** (2010), 1179–1188.

[83] K. R. Payne and D. F. Redaelli, *Primer on quasi-convex functions in non-linear potential theory*, arXiv:2303.14477v1, 2023.

[84] Y. A. Rubinstein and J. P. Solomon, *The degenerate special Lagrangian equation*, Adv. Math. **310** (2017), 889–939.

[85] Z. Slodkowski, *The Bremermann-Dirichlet problem for q-plurisubharmonic functions*, Ann. Scuola Norm. Sup. Pisa Cl. Sci. (4), **11** (1984), 303–326.

[86] G. Székelyhidi, V. Tosatti, and B. Weinkove, *Gauduchon metrics with prescribed volume form*, Acta Math. **219** (2017), 181–211.

[87] N. S. Trudinger, *The Dirichlet problem for the prescribed curvature equations*, Arch. Rational Mech. Anal. **111** (1990), 153–179.

[88] N. S. Trudinger and X.-J. Wang, *Hessian measures I*, Topol. Methods Nonlinear Anal. **10** (1997), 225–239.

[89] C. Villani, *Optimal transport. Old and new.*, Grundlehren der Mathematischen Wissenschaften 338, Springer, Berlin, 2009.

[90] A. Vitolo, *Maximum principles for viscosity solutuions of weakly elliptic equations*, 110–136, in Bruno Pini Mathematical Analysis Seminar 10, Università di Bologna, Bologna, Italy, 2019.

[91] S. T. Yau, *On the Ricci curvature of a compact Kähler manifold and the complex Monge-Ampère equation*, Comm. Pure Appl. Math. **31** (1978), 339–411.

Index

Printed in the USA
CPSIA information can be obtained
at www.ICGtesting.com
JSHW062052120823
46374JS00001B/1